Woods Hole Oceanographic Institution

Underwater camera with acoustic view-finder.

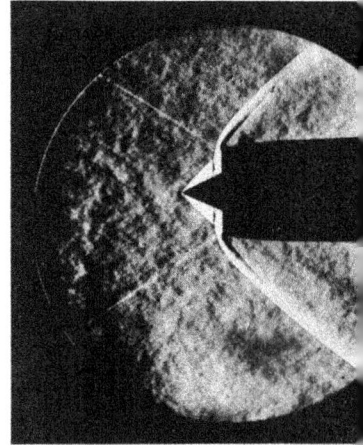

Supersonic shock waves around air inlet of ram-jet engine.

Raytheon Company

A portable fathometer with transducer.

The FIRST BOOK of
SOUND

The F-102A Delta Dagger, a supersonic aircraft

Convair

NACA Photo

Supersonic wind tunnel.

Solid sound. A plastic "loaf" model of the word "nine." This is a new technique enabling scientists to see sound as well as hear it. The model is constructed from "slices" of sound spectrograms recorded on special paper according to three dimensions - time, frequency, and energy.

The First Book of

SOUND

A Basic Guide to the Science of Acoustics

by David C. Knight

ILLUSTRATED WITH DRAWINGS AND PHOTOGRAPHS

Living Library Press
PO Box 16141
Bristol, Virginia 24209

FOR KARIN AND KARLA

The author's thanks to the following firms for their helpful cooperation with picture materials: Bell Telephone Laboratories, Inc.; Sonotone Corporation; Raytheon Manufacturing Company; Douglas Aircraft Company; U. S. Coast & Geodetic Survey; Woods Hole Oceanographic Institute; John-Manville Products; Westinghouse Electric Corporation; General Electric Company; Convair, A Division of General Dynamics Corporation; Sperry Products, Inc.; U.S. Air Force; Reeves Soundcraft Corp.; Telectrosonic Corp.; and National Advisory Committee for Aeronautics (NACA, which became NASA in 1958). Illustrations and diagrams not bearing a credit line were prepared by Peter Costanza.

Contents

Anechoic chamber, or "room of silence." Douglas Aircraft engineers designed this special chamber to duplicate the vast quiet of the stratosphere. The walls break up and absorb sound waves so that special acoustical effects can be studied. The anechoic chamber especially eliminates reverberations and echoes.

What We Mean by Sound

SOUND is something that we hear. It comes to our ears in many different ways. It may be pleasant, like the voice of a friend, or unpleasant, like the screech of a train's wheel on a railroad track. Some sounds are loud, and some are soft. Some are high, and some are low.

It is curious that most of the words we use to describe various kinds of sound are like the sounds themselves.

The ring of a bell. The bark of a dog. The shriek of a ship's whistle. The boom of a firecracker. The crack of a thunderbolt. All of these are sound.

Other sounds are represented by words like pop, squeak, hiss, bang, gurgle, rustle, crash. You can probably imagine how some of these words came to be. For example, water flowing in a brook gurgles. Mice squeak. Leaves rustle. Without much trouble you could make a long list of sound-words.

Sound is a form of energy that your ear, like a sensitive microphone, picks up from sources that are vibrating, or trembling, nearby or far away. The energy enters the marvelous system of your ear, which in turn delivers the energy to your brain, and you are then able to hear what we call sound.

Sound is extremely important in our lives. It is our chief means of communicating our thoughts, ideas, and wishes to other people. Stop a moment and think what your life would be like without sound. You would not be able to hear your teacher's voice in class, or your favorite TV program. You would not be able to hear the tune a band is playing in the park, or your best friend's voice on the telephone. In fact, the telephone would be of no further use at all.

If there were no sound, you would have to depend for communication on your remaining senses of touch, sight, smell, and taste.

But how does sound travel to our ears?

HOW SOUND COMES TO US

LET US pretend for a minute that you are standing on the surface of the moon. Some friends set off a charge of dynamite about a mile away. You can see the explosion, but you hear no report at all.

Why?

Because there is no air on the moon. You saw the flash of the explosion easily enough because light rays are able to travel through spaces where there is no air. But sound must have a *medium* to carry it along, or *transmit* it, from the place where it begins to your ear. A medium is simply the matter, or material, or substance, in which anything lives, or acts, or moves.

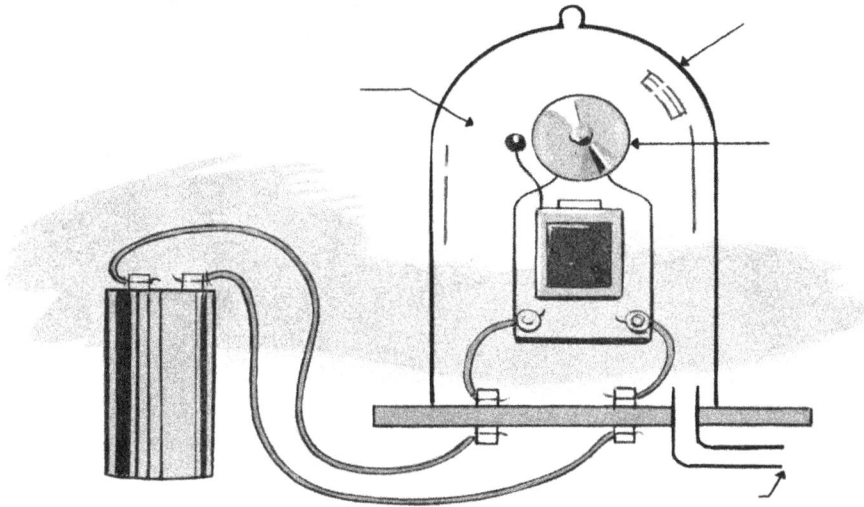

The bell-and-jar experiment.

When there is no substance—like air, wood, water, or anything else you can name—in a given space, that space is called a *vacuum*. Sound is unable to travel in a vacuum, although it can go through a solid object like a building.

In other words, sound needs a substance, or a medium, like air through which to go—or there is no sound.

To prove that sound cannot travel in a vacuum, you can perform an experiment. There are probably an electric bell and a large jar in your school science room. Set the bell ringing, place it under the jar, and start pumping the air out of the jar. Notice that the sound of the ringing bells becomes fainter and fainter as you pump. If you succeed in pumping most of the air out of the jar—thus creating a vacuum—you will not hear the bell at all. But let the air gradually creep back into the jar, and the ringing of the bell will grow louder and louder, because now the air is between your ear and the source of the sound.

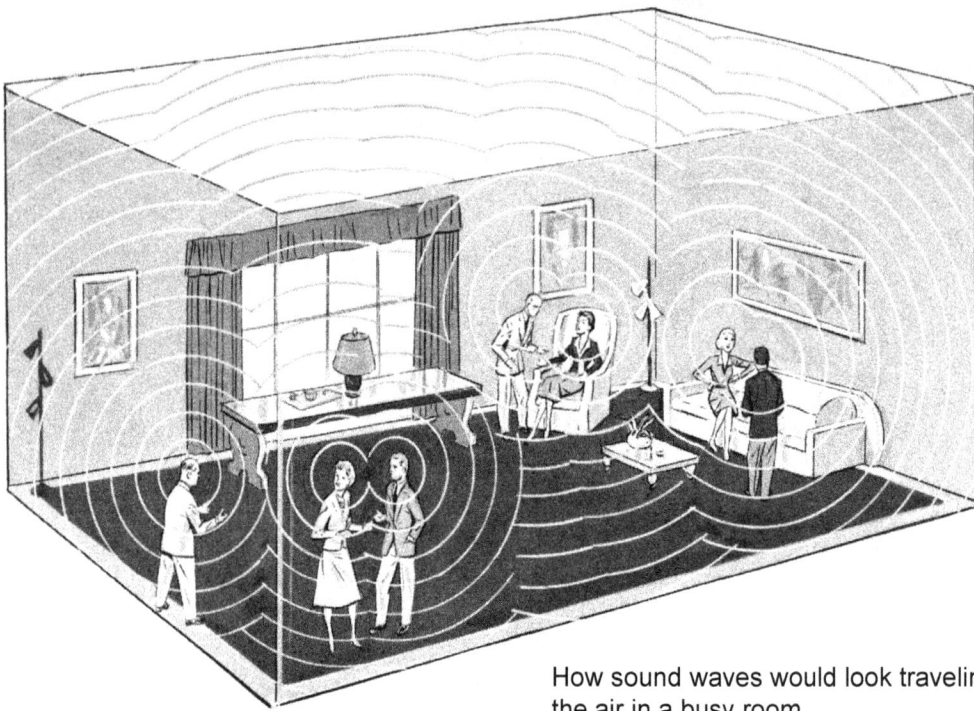

How sound waves would look traveling through the air in a busy room.

Air is the most common medium through which sound comes to us. Because air is so easily set in motion, hearing is possible. Without air, we should have to depend on other transmitters, or *conductors*, of sound, like water or steel. Can you imagine what it would be like if, every time you wanted to speak to someone, both of you had to stick your heads under water?

We—you and I, your parents, your teachers, all of us who inhabit the planet earth—live at the bottom of a thick ocean of air, which forms the earth's atmosphere. In one way or another, we are forever disturbing this air. It is difficult *not* to disturb it. When we walk about we push it aside. When we speak, we push it about by means of our vocal cords. Air, of course, is invisible. But if we could "see" a roomful of it—a room, let us say, in which many people were talking and doing things—we would see a very stormy sight indeed. The air would be disturbed by shakings and quiverings in all directions as people talked and moved. In many ways it would be like a huge block of jelly shot through with ever widening waves and ripples spreading out from their sources.

What are these sources? What happens to create sound?

WHAT CAUSES SOUND?

Every sound that we hear begins with a *vibration*. Vibrations even cause the sounds that we "hear" through our sense of touch—as we do when a friend taps a stick at one end and we feel the vibrations with our hand at the other.

What exactly is a vibration? It is *motion,* either backward and forward, or up and down.

You can see vibration in action by hitting a tuning fork with a rubber mallet. Notice the blurred appearance of the two prongs as they shake back and forth. Or take a rubber band and "twang" it between your thumb and forefinger and you will see the same thing.

Vibrations can be felt, too. Try this experiment: read the next sentence or two aloud. As you do so, place your fingers lightly in the vicinity of your Adam's apple. The tiny vibrations you feel are being made by your vocal cords. Or touch your alarm clock the next time it goes off, and feel the vibrating movements produced by the bell inside.

There are other ways, too, of showing how sound and vibration are related. Hit a tuning fork and hold it to the surface of a dish of water. Notice how the water is disturbed. Or take a Ping-pong ball and hang it on a piece of string. Hold a freshly hit tuning fork against the ball, and watch how the ball kicks and jumps.

Scientists have found out that an object must vibrate at least sixteen times a second before the human ear can hear it as sound. Likewise, most human ears cannot hear vibrations that are much faster than 20,000 times a second. If you took a broom, hung it on a nail, and started it swinging back and forth, there would not be enough vibrations produced to be *audible*, or hearable, to your ear. Or, if you blew the special kind of whistle made for calling dogs, you would not be able to hear it because there are too many vibrations. A dog, however, would be able to hear the whistle because his ear system can register higher vibrations than ours can.

The number of vibrations, or *oscillations*, that a sounding body makes *per second* is known as the *frequency* of that body. Any number of vibrations that a sounding body makes below 16 or much above 20,000 cannot be heard, or is *inaudible*, as far as man is concerned—although such vibrations *do* cause the eardrum to vibrate.

What Are the Different Kinds of Vibrations?

SOUNDING bodies usually fall into three main classes or groups of vibrations.

When you strike a tuning fork or the top of a table or anything else, its material vibrates at a frequency that depends on what the material is, as well as its size and shape. Such vibrations are "natural," and are known as *free vibrations*.

Everything in nature vibrates and has its own natural, or free, rate of vibration.

If a freely vibrating object, however, is left alone, the vibrations "die out." In other words, they will eventually grow weaker and stop. But if we take an alarm clock and start the alarm ringing, the

16

vibrations are renewed again and again. This is known as a *maintained vibration*—a vibration that is kept going. Another example would be that of a violin string which, when played with a bow, maintains its vibration until the bowing stops.

TUNING FORK
(Free vibration)

CLOCK-RADIO ALARM
(Maintained vibration)

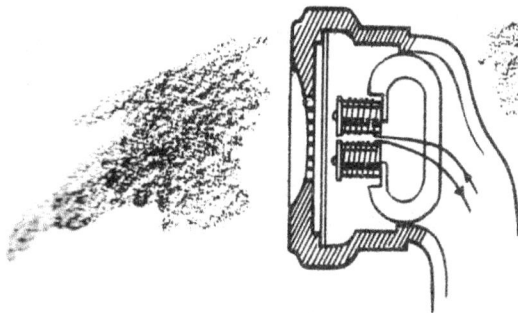

TELEPHONE DIAPHRAGM
(Forced vibration)

When a sounding body is made to vibrate at a frequency that is different from its natural one, this is a *forced vibration*. The diaphragm of a telephone receiver is made to shake with a forced vibration. So is the eardrum, which we shall consider next. Of course, the eardrum has its own frequency of free vibration, too. But it is curious that its free vibration is outside our range of hearing!

The Human Ear and How It Works

AS WE know, our daily lives would be difficult without sound for communication purposes. But suppose that, although there was plenty of sound about us, we had no ear systems to receive it. Again, we would be forced to depend on our remaining senses of taste, touch, smell, and sight, to do our communicating.

The human ear is a highly complex organ. Look at the cross-sectional illustration of it, and you will see that it is made up of three main parts: the *outer ear*, the *middle ear*, and the *inner ear*. In the outer ear, the part that sticks out from the head is called the *auricle*. The *canal*, in which wax forms as protection against foreign matter, leads into the skull.

The middle ear begins with the eardrum, sometimes called the *tympanum*. In the middle ear there are also three tiny bones called the *hammer*, the *anvil*, and the *stirrup*. These three tiny bones form a chain connecting the eardrum to the wall of the inner ear. The *Eustachian tube* is what connects the middle ear to the throat and serves to balance the air pressure on both sides of the eardrum.

The inner ear contains a spiral section called the *cochlea*. It is filled with a liquid, and is lined with many tiny nerve endings which are designed to receive sound vibrations.

When a vibrating wave of sound reaches the outer ear, it then passes through the canal to the eardrum. The eardrum then begins to vibrate, too. These vibrations are in turn passed along to the hammer, the anvil, and the stirrup, whose cross-piece is set into the shell-shaped cochlea. The fluid contained in the cochlea then acts on the many tiny nerve endings. Finally, these nerve endings transmit the vibrations to the brain by means of the *auditory nerve*.

The brain interprets these vibrations, or impulses, as one of the many varieties of sound.

A word of caution about your ears: never poke or probe carelessly into them. This can cause infection or serious injury to the sensitive eardrum. Be careful when diving from very high places.

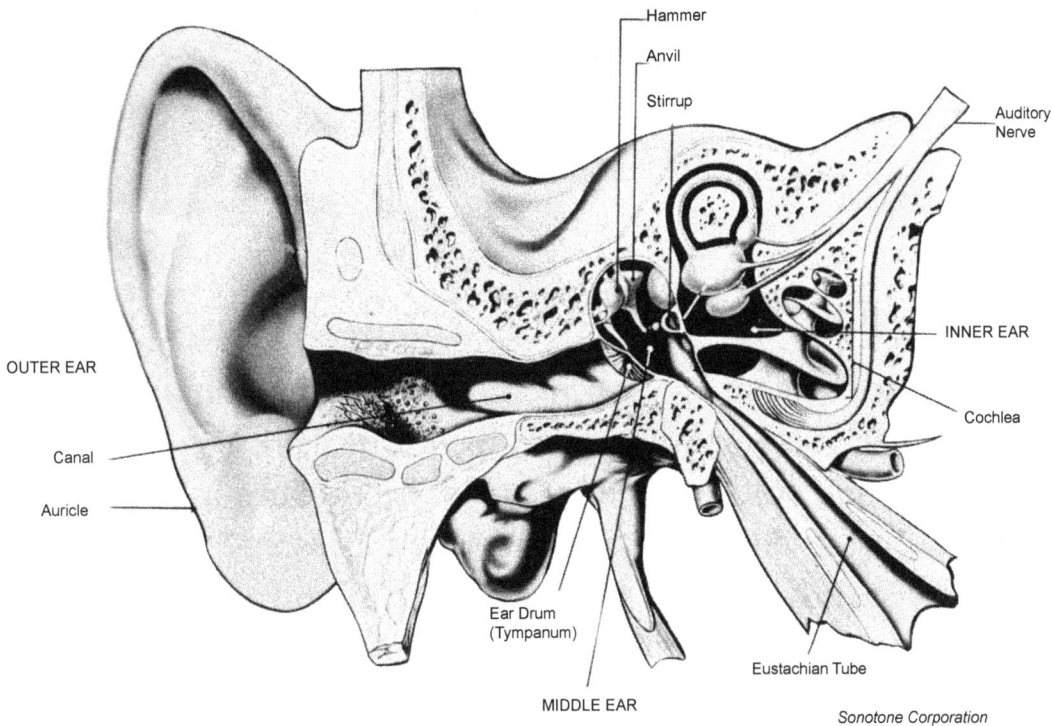

The human ear system

Remove excess water from the ears after swimming; it may be impure and could cause infection. Protect your ears as well as you can against sharp blows. Your ears are two of the most precious instruments you own. Take care of them well.

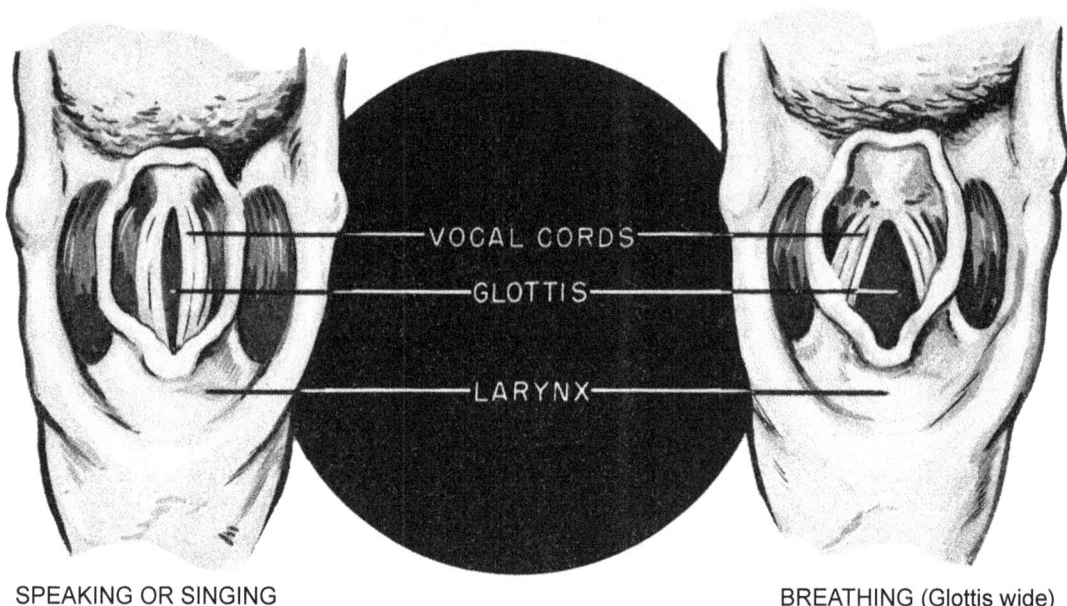

SPEAKING OR SINGING   BREATHING (Glottis wide)

Your Voice and How It Works

The human voice system is our principal means of communicating with one another. It is important to know how it works.

When the human voice produces sound, it produces it by means of vibrations, just as any other sounding body does.

In your upper throat, under your chin, is a slightly bulging portion, called your Adam's apple. Inside it is an apparatus called the *larynx*, which is a kind of boxlike chamber. Two tough membranes, called the *vocal cords*, are stretched across the larynx. Between the vocal cords is a narrow opening that is known as the *glottis*.

When you speak or sing, air is forced through the glottis, and your vocal cords vibrate and produce sound. If the vocal cords are close together (with the glottis opening small), your voice sounds are high. But when the glottis opening is wider, your voice sounds are lower because the vibration rate is lower. The shape of your mouth can also vary your speaking or singing voice.

A low bass singer's voice produces about 60 vibrations per second, while a high soprano's can vibrate as high as 1,200 times a second.

Sound Waves—What They Are and How They Act

WHEN a piano string or a tuning fork is struck so that it vibrates *audibly*, the air that surrounds it is disturbed by a series of tiny pushes.

These disturbances, which spread out in all directions from the vibrating source through the medium of air, are called *sound waves*. It is by means of these regularly spaced, or rhythmic, waves that sound energy is transmitted first to your outer ear, then through your middle ear, into your inner ear, and finally to your brain.

Although we usually think of sound waves as traveling through the medium of air, they can also be transmitted through other substances, or conductors such as glass, wood, steel, and water.

How Sound Waves Transmit Their Energy

SOUND waves behave very much like water waves.

Suppose you throw a stone into a perfectly quiet pond. The splash creates waves that spread out circularly in all directions.

Why?

Because the water level has been disturbed and is trying to get back to normal. But in doing so, the disturbance that was created only succeeds in spreading farther in the form of waves.

Soon a toy boat on the other side of the pond gets the "message" of the disturbance created by tossing in the stone. The boat now vibrates, or oscillates, too. In a sense, this is how the boat "pays for" the energy it has received from the original splash. Of course, the

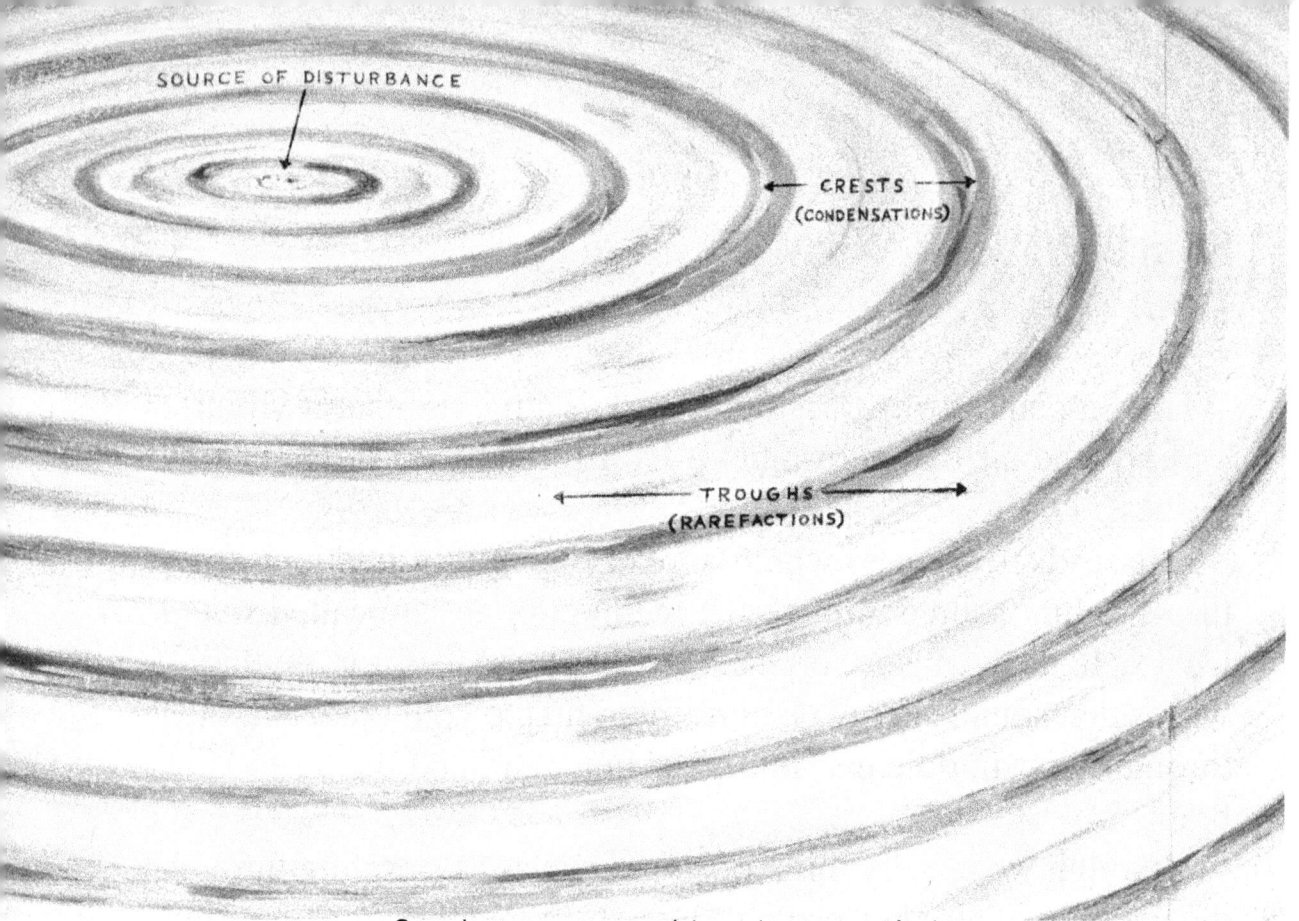

SOURCE OF DISTURBANCE

CRESTS
(CONDENSATIONS)

TROUGHS
(RAREFACTIONS)

Sound waves compared to water waves. A stone tossed into a pond creates waves with alternating crests and troughs, corresponding to condensations and rarefactions of sound waves.

farther away the waves get from the splash point, the weaker they become, just as sound waves become weaker as they travel from their source.

Sometimes these rhythmic waves, whether they are of water, sound, or electricity, are called *wave trains*. It is by means of them that the energy waves of a radio or TV signal are transmitted to a receiver, which is able to pick them up and replay them.

In the same way as the water waves transmitted energy to the toy boat, sound waves can transmit energy to your ear through air or some other elastic medium.

It is important to get a clear picture of how this is done. Notice the word "elastic" in the preceding paragraph. When we say that a medium or substance or conductor is elastic, we mean that it can be stretched or squeezed.

Suppose you take a tuning fork, tap it sharply with a rubber mallet, and hold it up to your ear. When one of the prongs of the fork vibrates *toward* your ear, particles of air called *molecules* are squeezed in the direction of your ear.

But when the prong snaps back *away* from your ear, the air molecules rush back, or *unsqueeze* themselves. Soon air molecules farther away, having been shoved by the first ones, transmit the original disturbance to your ear as a tiny dose of energy.

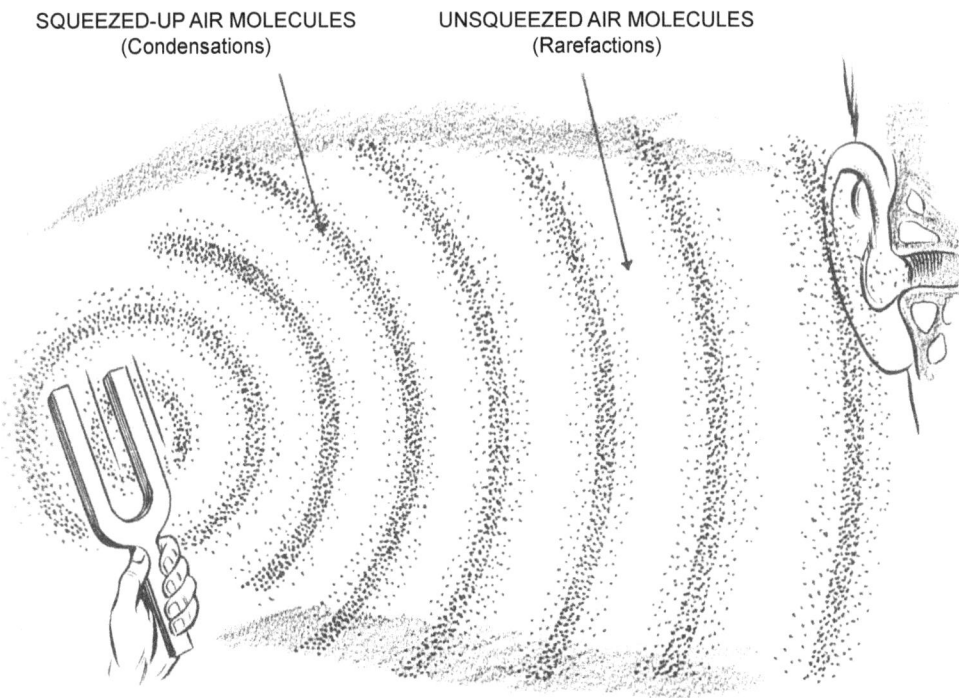

SQUEEZED-UP AIR MOLECULES
(Condensations)

UNSQUEEZED AIR MOLECULES
(Rarefactions)

The same basic thing would happen with a line of people, each with their hands on their neighbors' shoulders. If the person at the end of the line pushed or pulled on the shoulders of the person ahead of him, the same back-and-forth motion would be transmitted down the line from one person to the next.

What Kind of Waves Are Sound Waves?

SOUND WAVES are back-and-forth waves *only*. Physicists call this type of wave *longitudinal*, meaning lengthwise.

But there is another kind of wave called a *transverse* wave. It is important to recognize the difference between the two. Two simple experiments will show us the difference.

Take a rope about eight feet long and tie it to a post. With the rope held slack, shake it up and down until you have created wave motion along the whole length of rope.

Although the waves travel easily from your fingers to the post, every part of the rope is moving up and down, or *transversely* (perpendicularly), to the waves. No part of the rope is actually moving back-and-forth, or longitudinally.

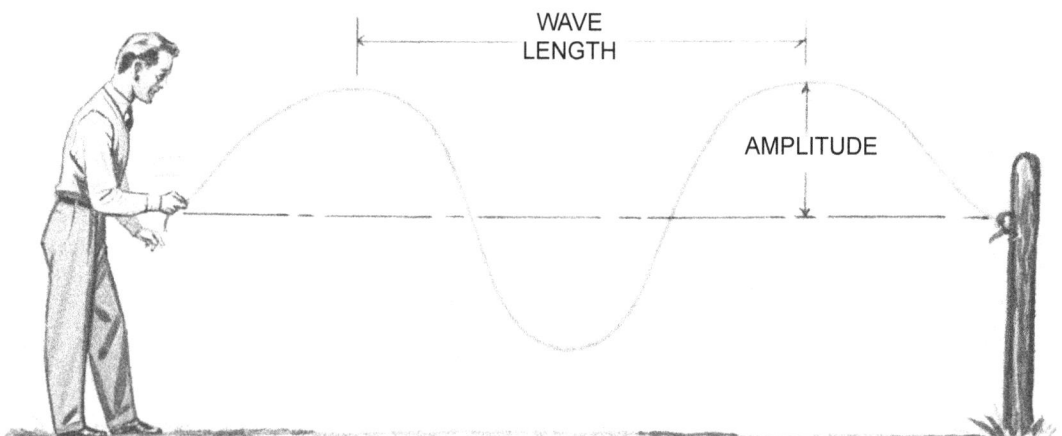

WAVE LENGTH

AMPLITUDE

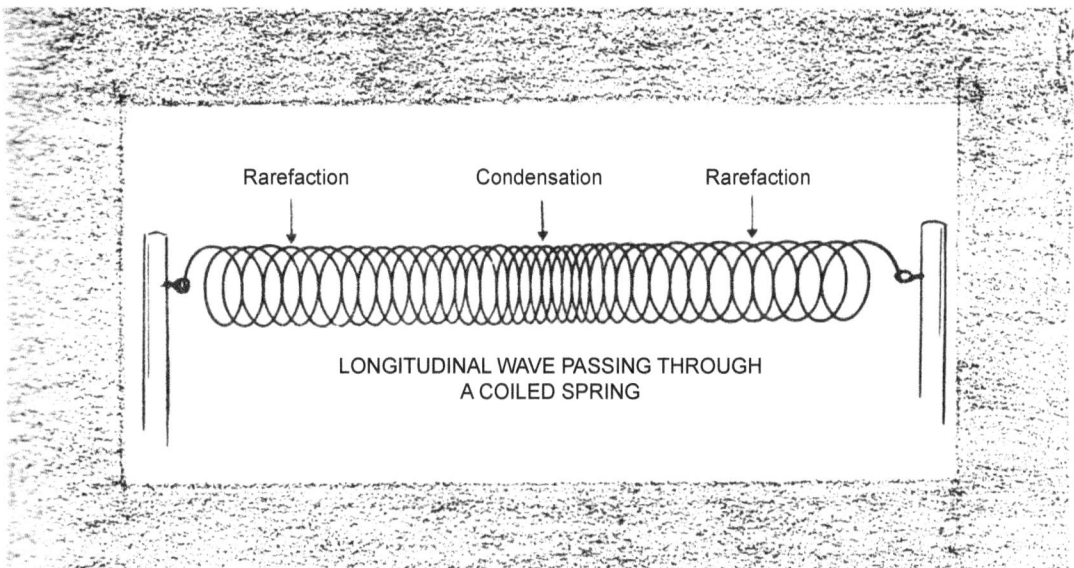

Rarefaction   Condensation   Rarefaction

LONGITUDINAL WAVE PASSING THROUGH
A COILED SPRING

The *wave length* of this type of wave is the distance between any two successive tops, or *crests*. The *amplitude* is the distance that any particle of the rope moves up and down, out of its usual place, as shown in the picture on the opposite page.

Light waves, as well as other kinds of electromagnetic radiation (such as radio waves), are transverse waves.

Now, take a long spring—one of the "Slinky" variety will do—and stretch it slightly between two posts. Take something thin, like a letter opener, and insert it between two coils in the middle of the spring. Jam some of the coils together in one direction and then pull the blade out quickly. Watch the resulting disturbance chase itself back and forth across the length of the spring.

What you have created in the spring is a *longitudinal* wave, sometimes called a *compression* wave. Longitudinal waves are back and forth waves, while transverse waves are up-and-down ones.

Sound waves are longitudinal waves.

The jammed-up places in the spring (like the squeezed-up air molecules between your ear and the tuning fork) are called *condensations*. And the stretched-out places in the spring (like the unsqueezed air molecules between your ear and the tuning fork) are called *rarefactions*.

Just as with transverse waves, we are able to measure two dimensions of longitudinal waves. The distance from one condensation to the next is called the *wave length* and, in the case of air, the distance that any given molecule is pushed back and forth is known as the *amplitude*.

The human ear can hear wave lengths ranging in length from about one inch to seventy feet. The faintest amplitude we can hear is about one-billionth of a centimeter, while the greatest amplitude we can hear is about one-thousandth of a centimeter. Sounds of greater amplitude than that would pain or even damage our ears.

Let us go back for a minute to the water waves created when the stone was tossed into the pond. What kind of waves are water waves? They are waves that are considered to be transverse because the water molecules move up, forward, down, and back, in a nearly circular path.

The Speed of Sound

IT always takes a certain amount of time for any sound wave to travel from one place to another through any medium. Even light rays, which do not need a medium through which to travel, are not transmitted all in an instant. Light waves, which travel through a vacuum at the incredible speed of 186,000 miles *per second* are of course much swifter than sound waves.

As a matter of fact, sound, when compared to the speed of other waves, is rather a slow traveler.

We know that sound waves must have some kind of medium through which to go before they can reach our ear systems. They cannot travel in a vacuum, as we saw from the bell-and-jar experiment.

Sound waves can, however, travel through liquids such as water, through solids such as glass or steel, and through gases such as air. Mediums such as glass and iron are good transmitters, or conductors, of sound waves. Some other mediums are poor conductors. For example, cloth and rubber are mediums that *absorb* or "soak up" sound waves.

Through "dry" air (air that is at 0° Centigrade), sound can travel at a speed of about 1,100 feet per second, or 700 miles per hour—about the speed of a bullet fired from a rifle.

As a general rule, we can say that most common liquids and solids transmit sound at a greater speed, or *velocity*, than air does. For example, sound travels about four times faster in water than in air. It also travels faster through the ground than in the air. Indians, by putting an ear to the ground, were able to hear the sound of approaching horses or buffaloes long before they could be heard through the air.

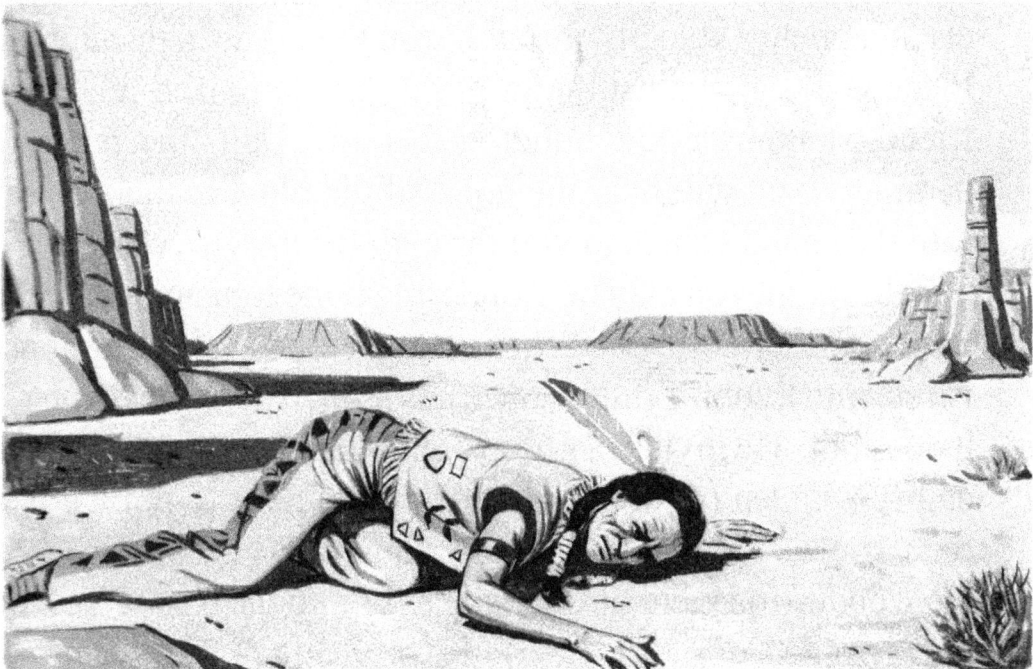

SOUND WAVES
TRAVELING THROUGH IRON

You can make an experiment yourself to prove that sound travels at a greater velocity through iron than it does through the air. Place your ear against the rail of a very long iron fence while a friend, standing a few hundred feet away, hits the fence with a hammer. You will hear the noise of the blow in the iron rail before the sound comes to you through the air.

At this point you may be wondering if the transmission of sound through, say, a solid body changes the sound at all. A famous British scientist, John Tyndall, once made an experiment to prove that it does not. He invited a group of people into a room to look at an empty wooden tray, supported from the floor only by a wooden rod. The wooden tray was mysteriously playing music. Actually the wooden rod went through the floor into the basement beneath,

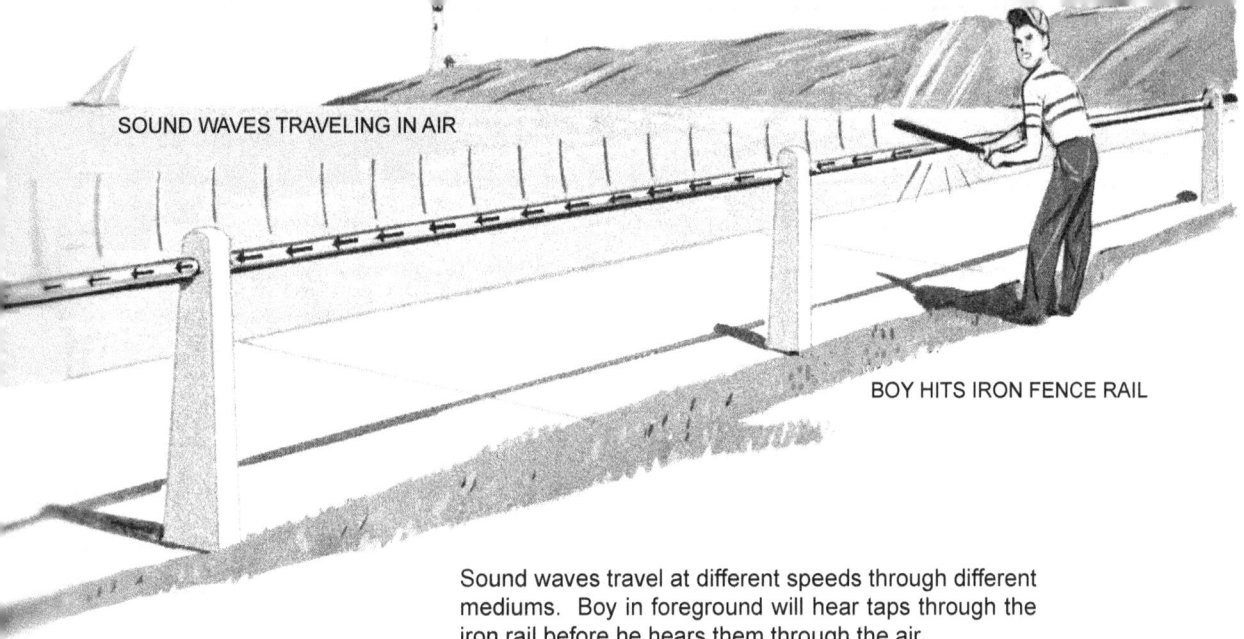

SOUND WAVES TRAVELING IN AIR

BOY HITS IRON FENCE RAIL

Sound waves travel at different speeds through different mediums. Boy in foreground will hear taps through the iron rail before he hears them through the air.

where the end of it was resting on a music box. The music itself, traveling up through the rod, had not changed at all. The tray merely acted as a sounding board for it.

In lead, sounds travel about 4,000 feet per second; in wood, about 11,000 feet per second; in glass, a little more than 16,000 feet per second; and in aluminum, just under 17,000 feet per second.

The ability of solid substances to transmit sound is a great blessing to people who have poor hearing. Completely deaf persons sometimes are able to hear music or other sounds by feeling vibrations through their fingers.

Thomas Edison, the great American inventor, for example, became quite deaf while he was still a young man. But he discovered an ingenious way to overcome his handicap. He explained once how he was able to listen to the phonograph, which he himself had invented. "I hear through my teeth and through my skull," Edison said. "Ordinarily I place my head against a phonograph, but if there is some faint sound that I don't quite catch this way, I bite into the wood and then I get it good and strong."

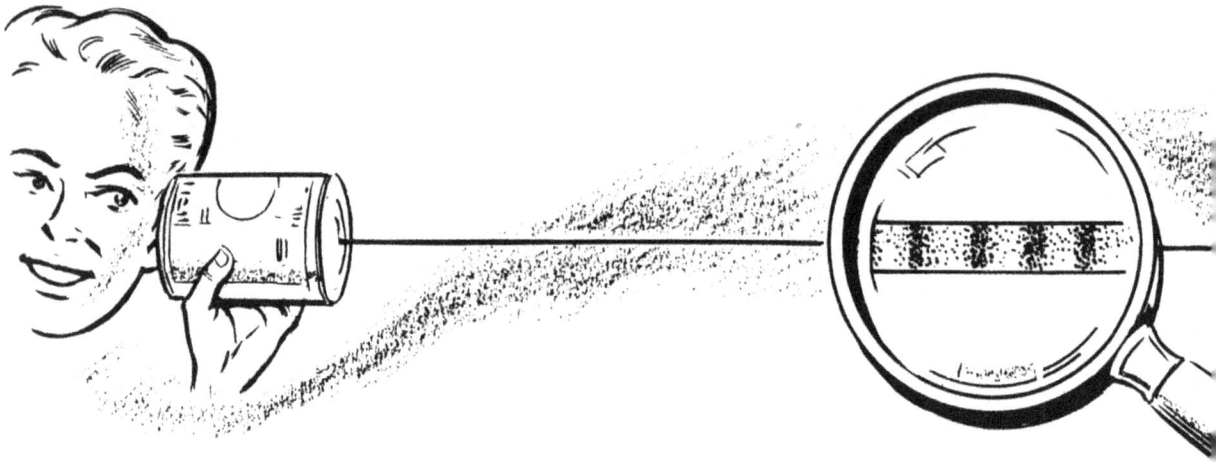

Another interesting example of how sound is conducted in a solid substance is illustrated by the "tin can telephone." This is fun to use and you can make one very easily. Just take two empty soup cans and stretch a very long wire or string between them. You can easily talk to a friend on the other end of the "line" by speaking into your can. Your friend will hear your voice vibrations in his own can which he uses as a receiver at his ear. The wire or string *conducts* your voice vibrations back and forth.

The more elastic a medium is (the greater its *elasticity*, in other words), the better and *faster* it can transmit sound waves. On the other hand, the greater the density (the amount of the medium in a given unit), the *slower* the speed of sound.

Why, then, you might ask, does sound travel through water much faster than through air, if water is denser? The answer is that while water is about eight hundred times as dense as air, it is many more than eight hundred times as *elastic*.

How Temperature Affects the Speed of Sound

Sound waves travel at different velocities as the temperature rises and falls.

HELLO! HOW ARE YOU?

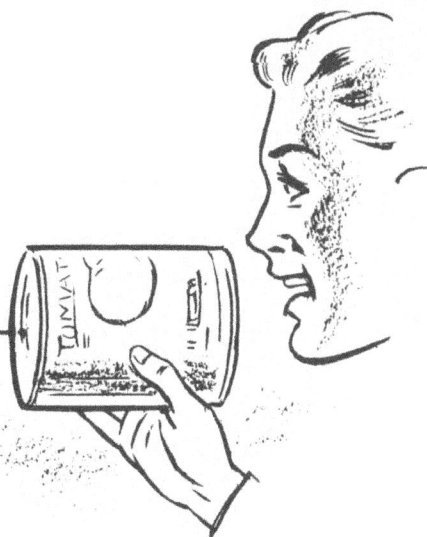

An important general rule to remember is that sound travels faster in air as the air gets warmer.

Why?

Because air expands when its temperature rises. This causes its density to grow less because there is less air in a given unit. You have already read that when the density of a medium grows greater, sound goes through the medium more slowly. The reverse is also true: when the density of a medium grows less, sound waves will be transmitted through it more quickly.

In the case of air, when it grows warmer its density lessens because of expansion. But because its elasticity remains about the same, sound waves will travel faster through it at a rate of increase of two feet per second with each centigrade degree rise in the temperature. Likewise, with each centigrade degree drop in the temperature of air, sound waves travel two feet slower. Thus, if you were standing on top of Mount Everest where it is considerably colder than at sea level, your voice would travel a few feet slower. However, if the temperature were the same on top of Mount Everest as it is at sea level, the speed of sound waves would be the same in both places.

We can say, then, that at any given temperature, all ordinary sound waves travel with the same speed through the same medum. And what a fortunate thing this is! Can you imagine how an orchestra would sound to us if the high notes of a piccolo reached our ear faster that the low ones of a bass violin?

The Three Characteristics of Sound

A "CHARACTERISTIC" is something that sets a person or a thing or a group apart from all others, and helps us recognize it. Take two friends, for example. Each one probably has at least three characteristics that make him different from the other one. Fred, perhaps, is nearly 6 feet tall, weighs 170 pounds, and is blond. Jim, on the other hand, is only 5 feet 6 inches tall, weighs only 140 pounds, and has a dark complexion. Jim and Fred are said to differ from each other by the characteristics of height, weight, and coloring.

The same thing can be said of sound, which has three main characteristics. These important identifying characteristics are *pitch*, *loudness*, and *quality*. Each one has something to do with the vibrations that sounding bodies produce.

Pitch

A *musical sound* or *tone* is usually pleasant to the ear because the vibrating source that produces it has a regular and even series of vibrations. Vibrations that are uneven or irregular do not usually cause *tone*, but *noise*. When you listen to a concert violinist drawing his bow over one of the strings of his fine instrument, you are hearing musical tone. But when your baby brother bangs a pot loudly on the kitchen floor, you are hearing noise.

Pitch, as well as the other characteristics of sound, is used to

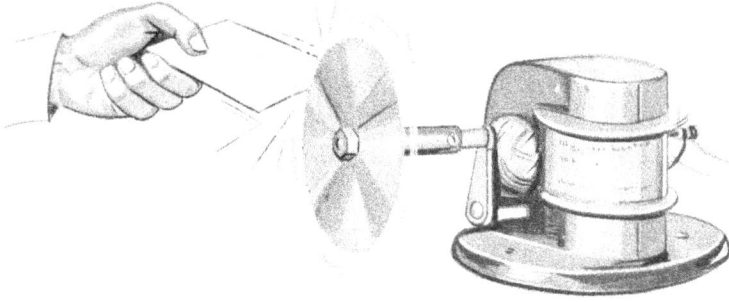
Savart's wheel

identify musical sounds rather than noise. *Pitch* refers to the lowness or highness of a sound. It is determined by the *frequency of vibration* (the number of vibrations a second) of the sound waves that strike the ear. The greater the frequency (the more vibrations a second), the higher the pitch. The lower the frequency (the fewer vibrations a second), the lower the pitch.

So that you may understand pitch better, try this experiment. Ask your science teacher for a *Savart's wheel*. This is simply a small wheel with regularly spaced teeth on its outer rim. It is attached to an electric motor that causes it to spin rapidly.

If you hold an ordinary playing card against the toothed wheel as it spins, the card will vibrate and give out a musical sound. The faster you rotate the wheel, the higher the *pitch* of the sound will be. When the wheel is slowed down, the pitch is lowered.

In the case of the Savart's wheel, we can say that each time one of the teeth strikes the card, one vibration is produced. The frequency vibration of the card, which determines its pitch, can be found by multiplying the number of teeth on the wheel by the number of revolutions it makes a second.

Thus, if the wheel has one hundred teeth and it rotates at three revolutions a second: Frequency = number of teeth x revolutions per second = 100 x 3 = 300.

The "highness" or "lowness" of a sound depends more or less on each person who hears it. A note that sounds "very high" to you might not sound "very high" to a friend at all, and his idea of what is "low" would be different from yours, too.

In general, we can say that pitch depends on frequency because every time frequency changes, pitch changes, too. If the frequency of a musically sounding body doubles, however, we cannot say that the pitch doubles also. It is true that the pitch is higher, but who is to judge *how much* higher? In the same way, you would say that your lima beans are saltier (but not *twice as salty*) if you put two pinches of salt on them instead of one.

In the making of musical sounds, musical instruments vibrate at different frequencies and give off *notes* of higher or lower pitch, which can be written down on *musical scales*. When your music teacher sounds the "key" on her pitch pipe, she is giving you the definite frequency, or pitch, that begins the song she wants you to sing. When a policeman blows his whistle in the midst of heavy traffic, you can hear it clearly because its high pitch can easily be heard over the lower frequencies of the automobiles' motors.

Police cars and ambulances also rely heavily on a pitch-changing device called a siren for a warning. You can learn exactly how it works by experimenting with the *siren disk* in your school science laboratory. This siren disk is a flat plate with regularly spaced holes in it. You will notice that as the circular rows of holes go toward the center of the disk, each row has fewer holes.

The disk is rotated by a small electric motor. As it goes round, use a small hose as an air-jet, and blow air on the innermost row of holes. A different note will be heard for each row of holes. You will notice, too, that as the air is blown on the rows containing more holes, the pitch rises. This is because the frequency increases as the number of holes increases.

Since the holes in the disk are evenly spaced, musical notes having varying pitch are produced. But if one of the rows had holes that were not evenly spaced, the result would be *noise*, assuming of course that the air stream and the rotation of the disk did not change from time to time.

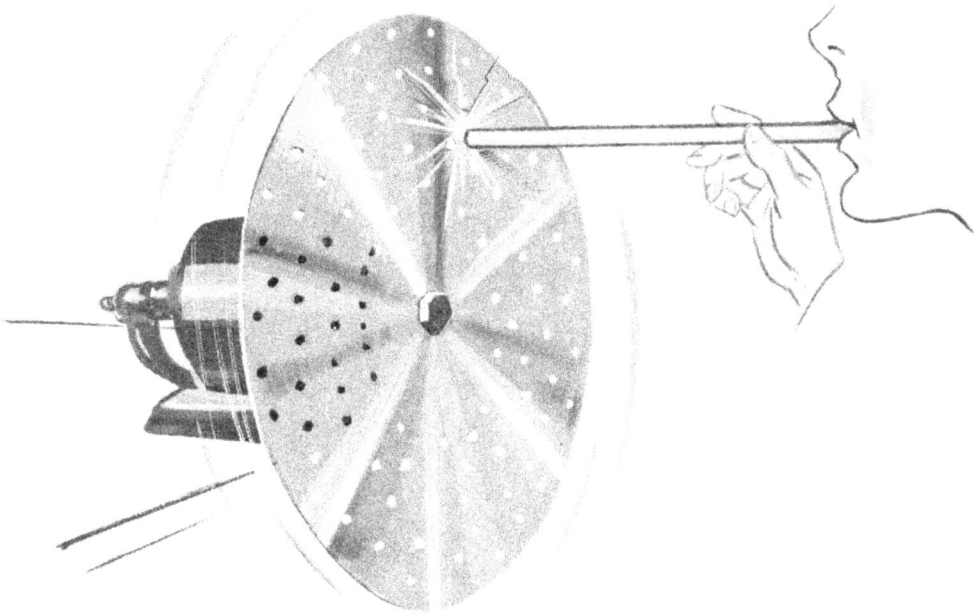

A siren disk

Suppose you are standing on a street corner and you hear the siren of a rapidly approaching squad car. For our purposes we will assume that the siren is giving out sound waves of one constant, steady pitch. Since the car is coming toward you and disturbing the air, more waves will strike your ear than if the car were standing still. This is the same as saying that there is an increase in frequency. Consequently, the pitch of the siren rises. But when the car passes by and goes away from you the pitch is lowered. Named

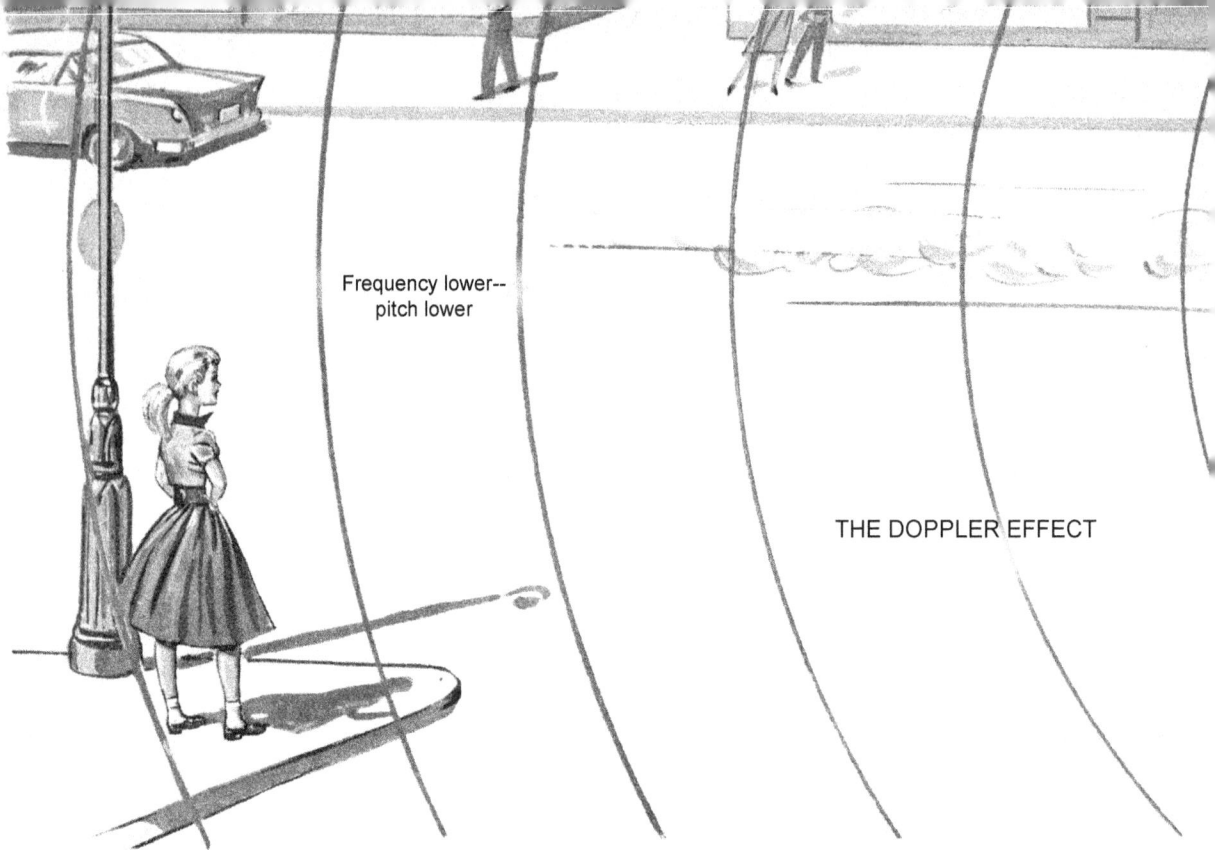

Frequency lower--
pitch lower

THE DOPPLER EFFECT

after its discoverer, an Austrian physicist, this is known as the *Doppler effect*. This effect is defined as the change in the pitch of a sound due to the rapid change in the distance between an observer and a sounding body's source.

Loudness

THE second characteristic of sound is its *loudness*. A cannon shot roars out its sound waves. A pin dropping on the floor produces sound waves that are so faint they are almost inaudible. The loudness of a sound is determined by *the force with which its waves strike the ear system.*

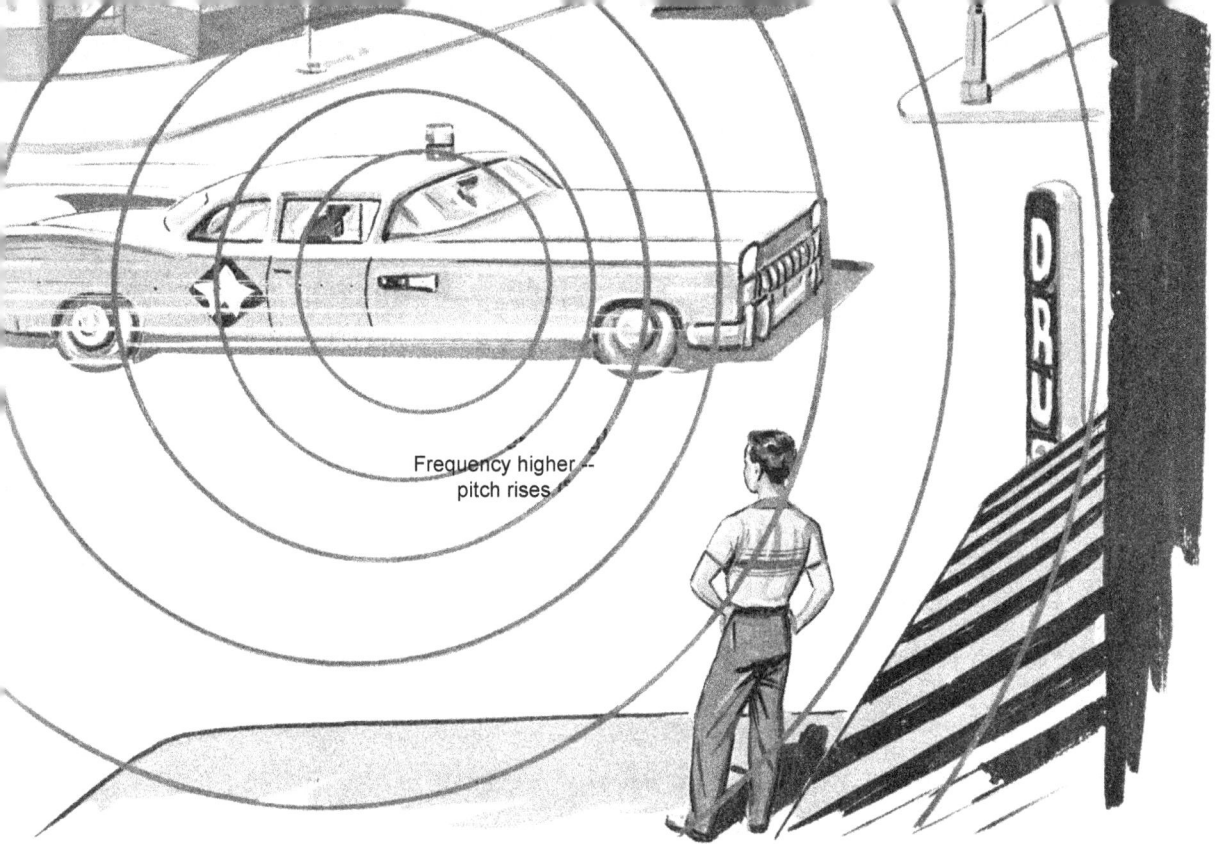

Frequency higher --
pitch rises

Naturally, the closer you happen to be to the sounding body's source, the louder the sound will be. A soldier who fires a large artillery piece is so close to this terrific sound source that he is forced to deaden it by stuffing his ears with cotton. But an infantryman several miles away will probably hear only a very faint report.

Loudness is much like pitch in one way. Each person has his own idea of what is loud, and loudness cannot be measured. But sound engineers have found out that the *intensity,* or the *rate* at which sound waves produced by vibration do work on our ears, *can* be measured.

The unit that expresses these various intensity levels is called the

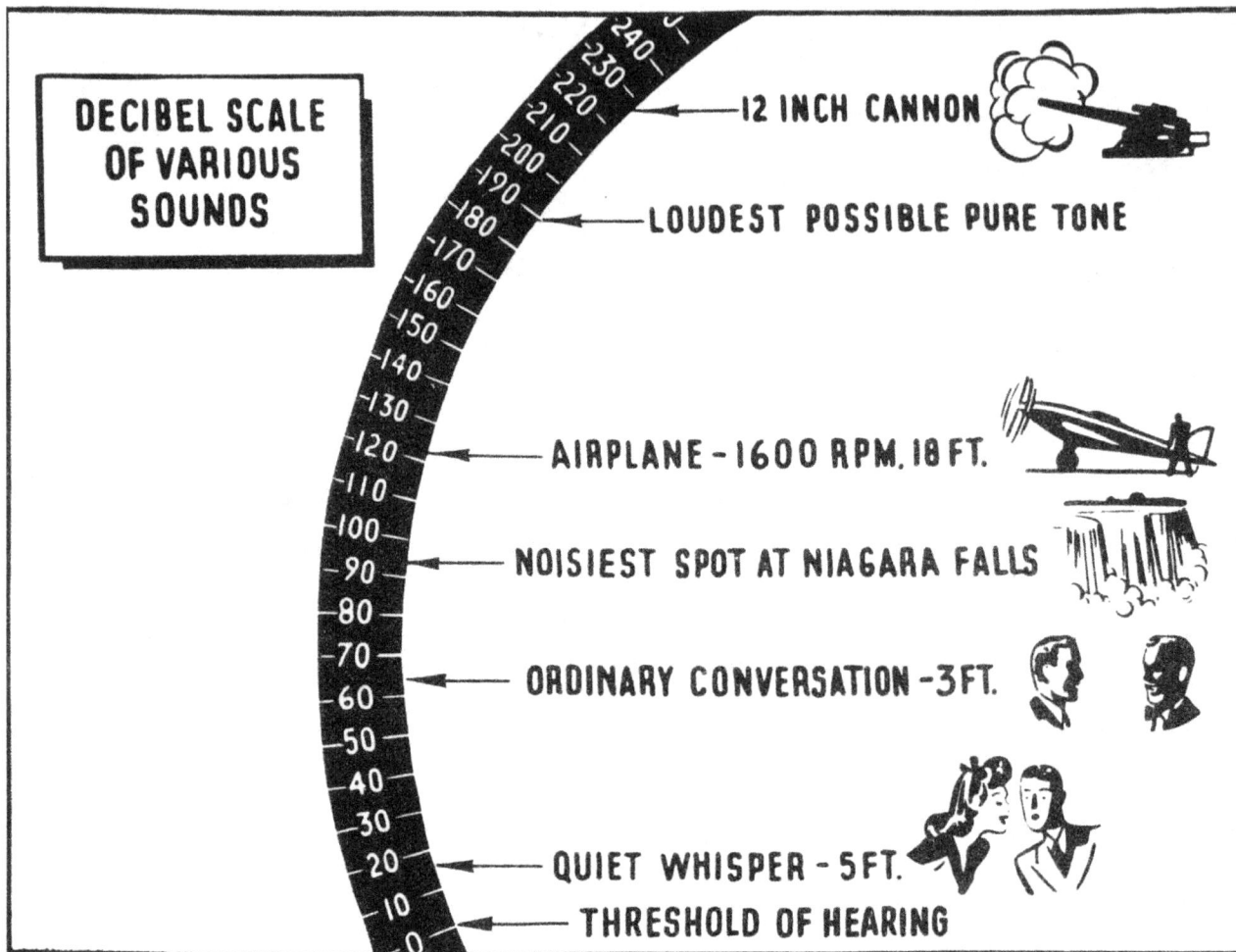

DECIBEL SCALE OF VARIOUS SOUNDS

12 INCH CANNON

LOUDEST POSSIBLE PURE TONE

AIRPLANE - 1600 RPM. 18 FT.

NOISIEST SPOT AT NIAGARA FALLS

ORDINARY CONVERSATION - 3 FT.

QUIET WHISPER - 5 FT.

THRESHOLD OF HEARING

decibel. Look at the decibel scale of sounds on this page. About the slightest *difference* in intensity that the human ear can detect is a change of 1 decibel. Starting at the *threshold of hearing* (the point where a sound can be heard at all), the intensity level of a whispering voice is about 20 decibels; the level of normal conversation, about 65 decibels; the level of traffic in a busy street, about 80 decibels. The noisiest spot at Niagara Falls registers at about 95 decibels. If we hear sound much above the 120 decibel range, it becomes painful to our ears.

Quality

IN ADDITION to the pitch and loudness of sound, there is also a third characteristic—*quality*.

Suppose you entered a darkened room and a musician started playing different instruments one after another. And suppose that on each of them he sounded just one note of the same pitch and loudness. Would you recognize the bass violin? The cornet? The trombone? Or perhaps the musician's voice itself if he hummed the same note? The chances are that you would, because each has its own peculiar quality.

If we were aware of the loudness and pitch of sounds only, it would be hard to tell them apart, since those of the same frequency would sound pretty much alike.

Scientists who have closely studied sound quality tell us that any sounding body is capable of vibrating at several *different* frequencies *at the same time*. Because a sounding body can do this, it has its own particular quality.

When the string of a musical instrument vibrates, it can do so all at once in a back-and-forth motion, or it can vibrate in parts to make *overtones*.

You can make a simple experiment to show what overtones are. Get a length of rubber tubing about twenty feet long and tie one end of it to a doorknob. With the other end of the tubing in your hand, stretch it fairly tightly and move your hand up and down slowly. Soon you will get the tubing to move up and down as a single unit or wave. Now, pretend that your rubber tubing is a very long violin string which is producing a note. That note would be the violin string's lowest one because it is vibrating in just one wave along its whole length: such a note is called a *fundamental*.

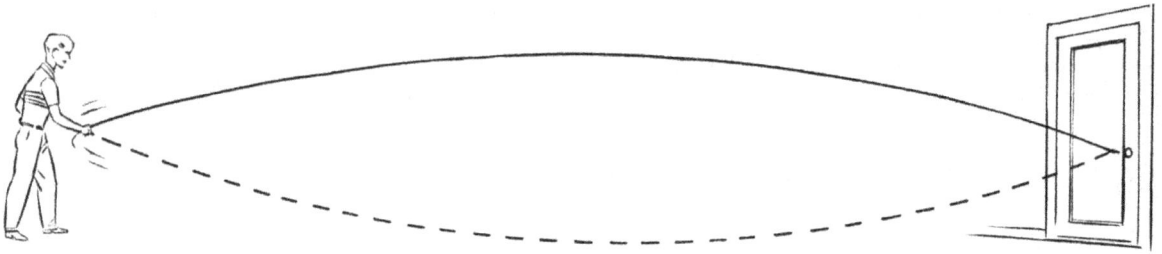

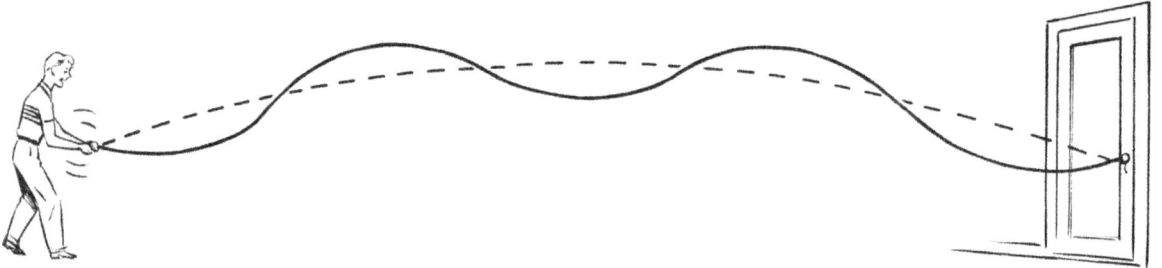

The boy above is producing the *fundamental* in his rubber tubing by making it vibrate in one wave. The boy below has produced *overtones* in his tubing at the same time it is swinging up and down in the fundamental.

Now, being careful to move the tubing up and down only—not in a circle—move your arms faster. With practice, you can make the tubing vibrate in two sections, then in three, and four, and five. These are the tubing's overtones.

Pretend again that the tubing is a very long violin string. If it is vibrating in two sections, the notes made by the two halves are called the string's *first overtone* (sometimes called a *second harmonic*). If the string vibrates in thirds, the string is producing its *second overtone* (or *third harmonic*). The places in the string (or tubing) where there is little or no vibration—that is, where there is little or no up and down movement—are called *nodes*.

Good violinists can bow a violin string to make many overtones, depending on where their bows are placed. Sometimes their strings vibrate as a whole unit with overtones *added on* to the single fundamental

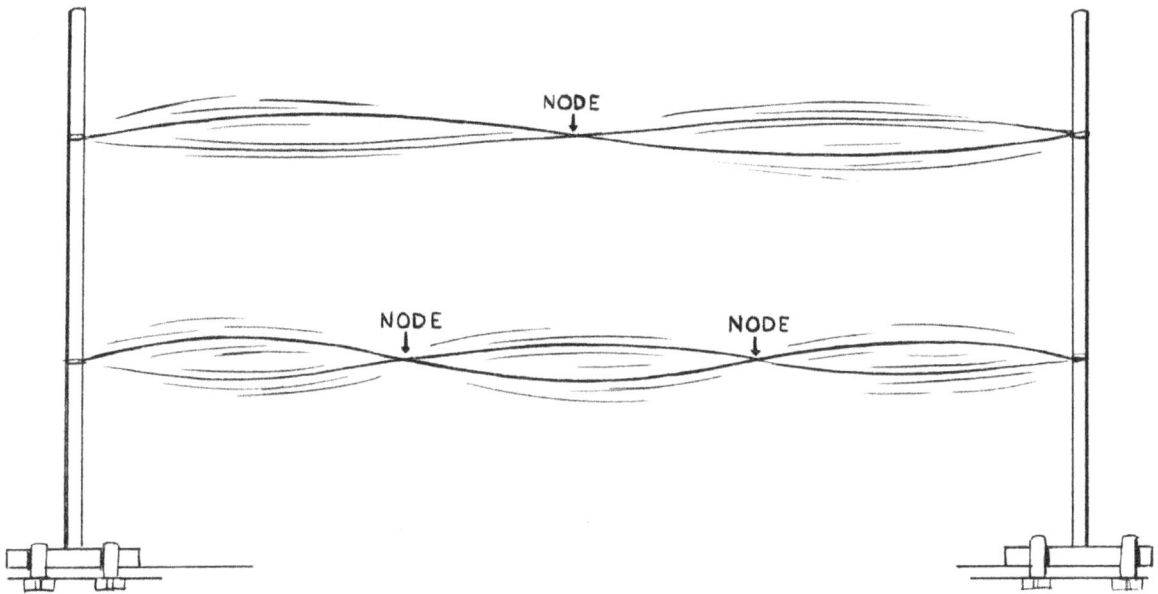

The violin string above is being made to vibrate in halves, producing its *first overtone*; the string below is vibrating in thirds, producing its *second overtone*. Although not shown, the strings are vibrating in their *fundamentals* as well.

vibration. If you practice some more with the rubber tubing, you can make it swing up and down in one whole unit and at the same time make overtones in the individual parts of it.

The third characteristic of sound, *quality*, is determined by the number and intensity of the overtones that are produced by vibrations in the divisions of a sounding body. If we take a musical string and make it vibrate as a whole 256 times a second, we hear its fundamental. But if we make the halves of the same string each vibrate twice as fast, or 512 times a second, the resulting sound is made up of two sounds—the fundamental and an overtone—since both the *halves* of the string and the whole string are vibrating *simultaneously*. Although the fundamental may be stronger, both sounds are still heard. Usually, the fundamental tone determines the main pitch and loudness of, say, a musical instrument; but the overtones

41

can also be heard. If these *harmonize*, or blend, pleasingly, the quality is pleasant to our ears. If the harmony, or blending, is poor, the sound is harsh and unpleasant.

No two instruments have ever been made that have exactly the same quality, although they may have the same pitch and loudness. In a violin, this is due not only to the adjustment of the strings, but also to the particular construction of the sounding box attached to it. The difference between a $15 violin and a $15,000 one is that the latter produces sweeter, mellower musical tones. Somehow, by expert craftsmanship and musical "ear", the man who made the $15,000 instrument was able to shape the materials so that the fundamentals and overtones for every note blended perfectly. An eighteenth century Italian named Antonio Stradivari became famous for his skill in making fine-sounding violins, and today a genuine "Stradivarius" sells for many thousands of dollars.

Any two sets of vocal cords are no more exactly alike in quality than are two musical instruments. We can tell one friend's voice from another's because we have learned to tell the differences in their vocal-cord vibrations. One of the reasons an opera singer can give a pleasing performance is that she is the fortunate owner of a set of vocal cords whose fundamental and overtone vibrations blend in a highly pleasing manner.

Modern science has perfected many new methods of analyzing complicated sound waves. Shown on the opposite page are *spectrograms* of a number of different voice sounds. These are made by a late-model device called a *sound spectrograph*. This is a useful laboratory tool for determining the nature of the time and energy distributions of sound.

SONG — TRAINED SOPRANO VOICE

CRY — BABY

LAUGH

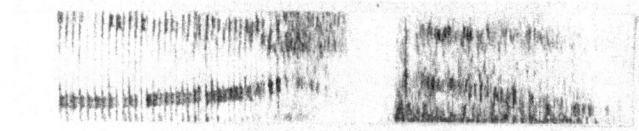

SNORE

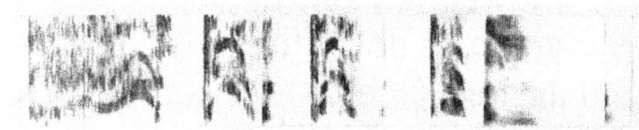

COUGH

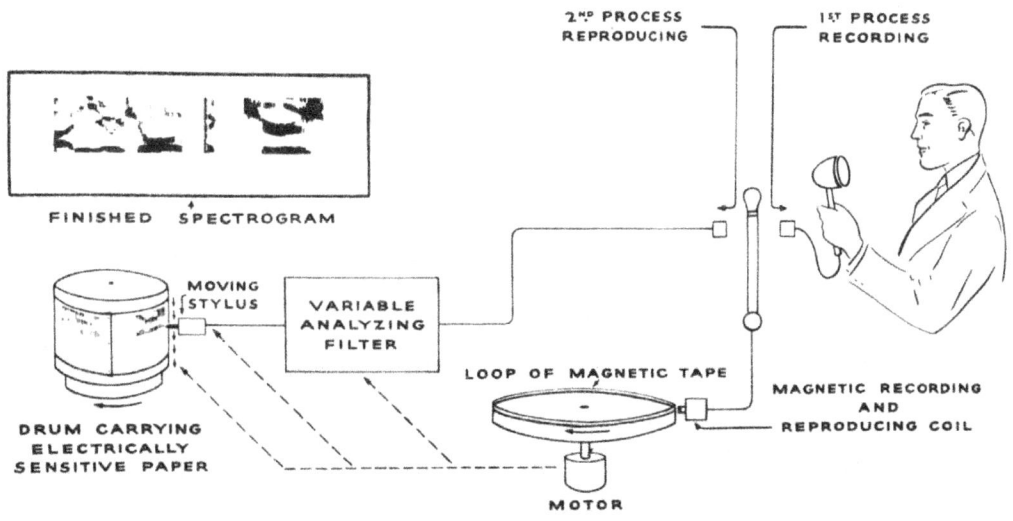

Spectograms of familiar voice sounds made by a sound spectograph. Speech is first recorded on magnetic tape when a switch is pulled to the right. When the switch is at left, the loop of tape is played back once for each of the pitch groups. The moving stylus indicates the loudness of the sound in each pitch group.

Harmony and Musical Instruments

MUSICAL sounds are far more pleasant than noise because they result from vibrations that are regularly spaced. These pleasing sounds are something like the steps of a stairway. In the music we know best, each group of them has eight steps arranged according to a definite plan; they form a scale. The eight steps are known as an *octave*. You will probably recall from your music class that the eight notes, *do, re, mi, fa, sol, la ti, do* form the *diatonic scale*, which is generally in use by Americans and Europeans today. Other scales having fewer or more than the eight steps, except for the chromatic scale which has twelve notes, sound peculiar to our ears, unless we are accustomed to listening to them.

When a sound pattern allows the singing voice or a musical instrument to go up or down the musical steps in regular order, we usually call the result a *tune* or a *melody*. And many musical tones sounded together can be blended into an agreeable combination of sounds called *harmony*. When you listen to a barber shop quartet "harmonizing", you are hearing the results of an agreeable combination of human voices moving up and down the diatonic scale in rhythm.

Musical instruments are made to produce sounds which are regular in pitch and agreeable in quality to our ears. They are generally divided into three main groups: *wind, stringed, and percussion.*

The notes that are made by wind instruments are produced by vibrating columns of air within the instruments. Some of the wind instruments are the organ, the trumpet, the flute, and the trombone.

You can easily make a trombone of your own by dipping a length of half-inch glass tubing in a tall glass of water. As you move the tube up and down in the glass, blow across the top of it with a "spreader" such as is used to spread the flame of a Bunsen burner.

Various famililar musical instruments. Which belong to the *wind* group? Which to the *strings*? Which to the *percussion*?

Notes of varying pitch will be produced as the vibrating air column is shortened or lengthened by the tube's depth in the glass.

Stringed instruments make use of a stretched wire string or animal gut, usually mounted over a sounding box. The strings are caused to vibrate by plucking or bowing. Some of the stringed instruments are the harp, the violin, and the viola.

Percussion instruments produce their musical sounds when they are struck by the hand, a stick, or a covered hammer. Some of them are the xylophone, the cymbals, bells, and the drum. The piano is considered a percussion instrument rather than a stringed one because it has felt-covered hammers that *strike* the strings which are inside the instrument.

Reflected Sounds — Echoes

AT SOME TIME or another, we have all been amused by the echoes we hear when we shout "Hello!" at a nearby hill, a cliffside, a wall, or a steep riverbank. After a short pause, back comes our voice with the last part of whatever we shouted. What has happened?

The air disturbance caused by the voice has been transmitted outward until it reaches the hillside. From there it bounces back again, and we hear the same disturbance that we ourselves created a second or two before.

In other words, sound waves can be reflected, just as light rays can be reflected from a mirror. In general, the smoother the reflecting surface, the better the echoes that the sound waves produce, because the waves are less distorted.

Since we know the speed of sound in various mediums, we can calculate with fair accuracy how soon echoes will return. For example, if you were standing 550 feet away from a steep cliff, the echo of your voice would return to you in exactly one second.

Why?

Because it would take the sound disturbance a half-second to get to the cliff and another half-second to travel back, since sound travels 1,100 feet per second in air.

There are many practical uses for echoes. Navigators at sea frequently use echoes to warn them of possible danger. The ship's captain sounds his foghorn if he is sailing through a fog or if he suspects that rocky cliffs are near at hand. The echo of the horn's sound tells him whether or not he is a safe distance away from the danger. Likewise, foghorn echoes are used in Arctic waters to signal approaching peril from icebergs. In each case, it is the amount of time between the foghorn's blast and the echo that tells how far away the ship is from possible danger.

The depth of the ocean can be charted by the use of echoes. Short sound waves are sent out from a ship by means of an oscillator, and are bounced off the ocean bottom. When these waves return they are registered and depth is shown by an instrument called a *fathometer*. When a number of such readings have been gathered, charts of the ocean floor can then be drawn.

U.S. Coast & Geodetic Survey

A recorder and an operator using the fathometer, or sonic depth finder, in a surveying launch.

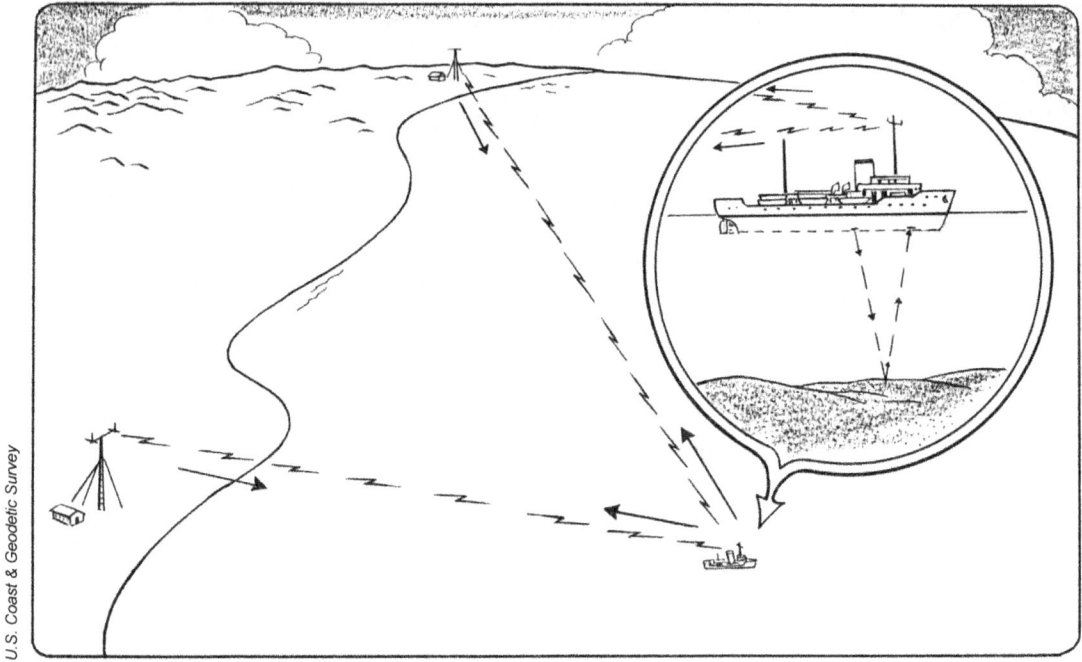

U.S. Coast & Geodetic Survey

Hydrographic surveying. Two methods of obtaining depths on nautical charts, and their precise positions. Depths are determined with a fathometer which measures time requried for an echo to travel to the sea bottom and back to the ship. These are made 4 times a second as the ship goes back and forth. Depth positions are fixed by electronic methods from two or more ground stations.

A *seamount* (mountain beneath the sea) 3,360 feet high in the Caribbean Sea. Recording was made with Woods Hole Precision Echo Sounder Recorder aboard the Research Vessel *Crawford*.

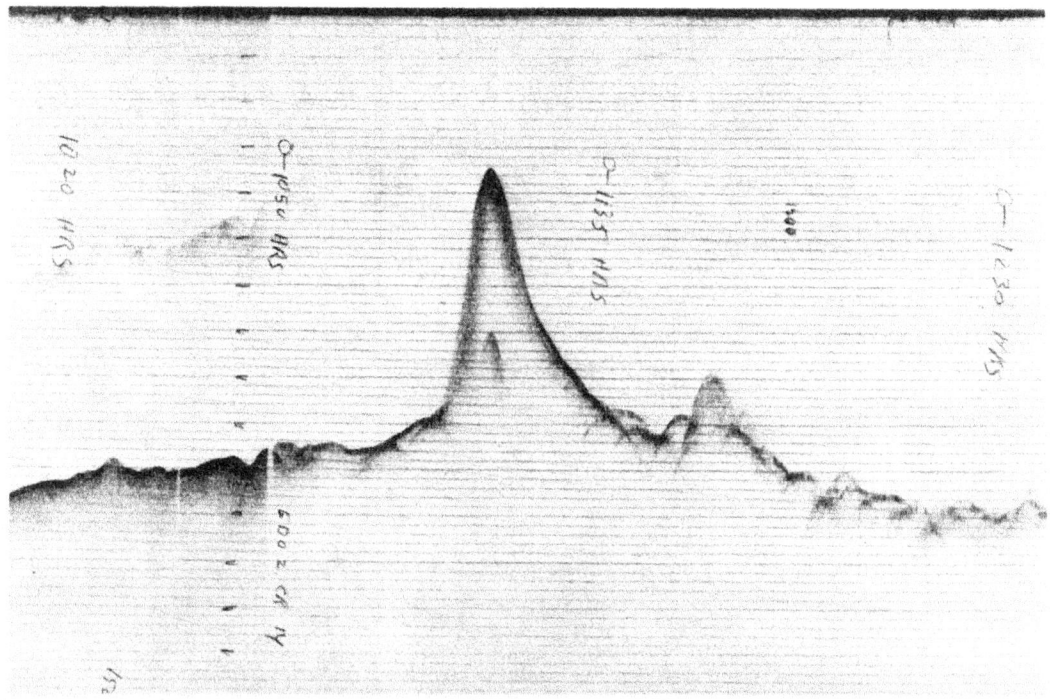

Woods Hole Oceanographic Institution

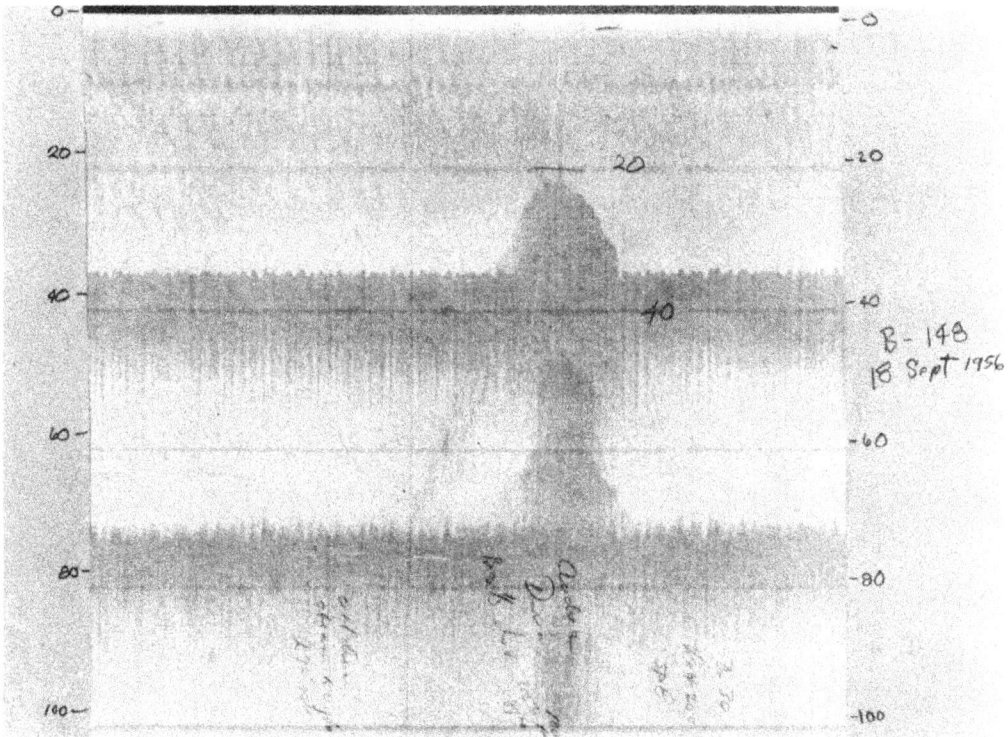

Woods Hole Oceanographic Insitution

Bathygram of the *Andrea Doria* sunk off the New England coast in about 230 feet of water. The ship is lying on her side. The lower recording is a double echo made on second "sweep" of the sounder recorder. Second echoes are common in such shallow water.

Typical reflection "shot" at sea. A small explosive is set off at the surface to determine the thickness of sediment on the ocean floor. Two sets of sounds reach the *hydrophone*, or water-listening device: one reflected from the ocean bottom, the other from the sub-bottom.

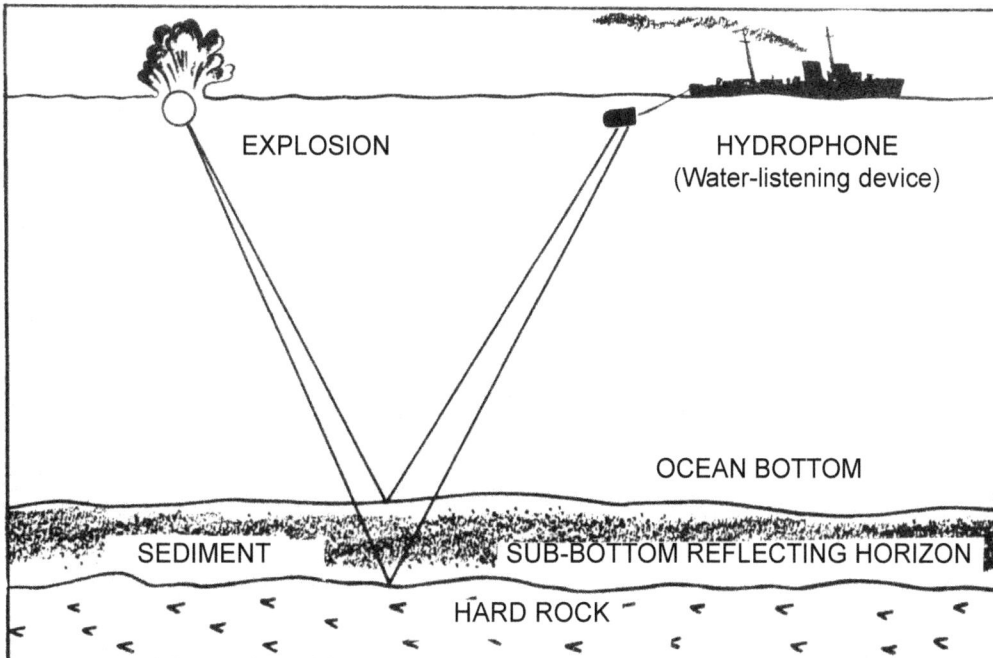

Woods Hole Oceanographic Insitution

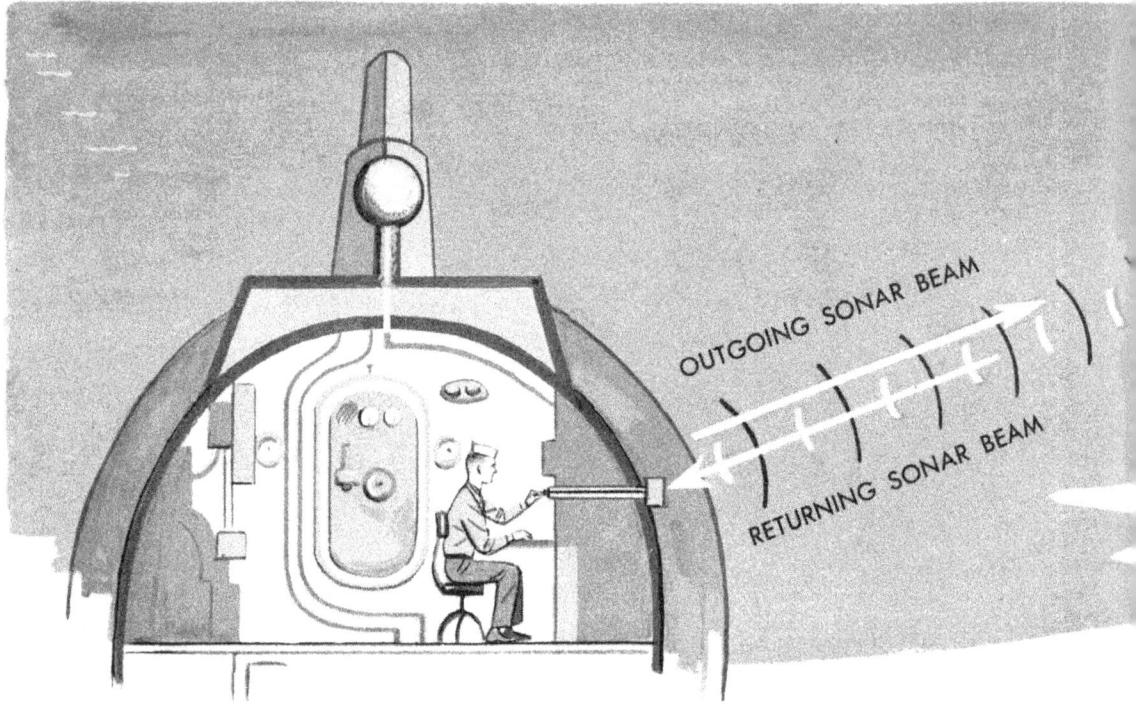

OUTGOING SONAR BEAM

RETURNING SONAR BEAM

The United States Navy also makes practical use of reflecting sound waves in an instrument called sonar. The name comes from a combination of the letters from the words *Sound Navigation And Ranging*. From a rather simple water-listening device, sonar has developed into a very delicate piece of equipment which can tell its operator when ships are near, in what direction they are located, and exactly how close they are. Sonar, of course, is especially valuable to submarine crews who must stay submerged for long periods of time while hiding from enemy surface ships.

Sonar is essentially high-frequency sound waves that are "beamed," or sent out, under water, usually in all directions so that they will detect some target. Any solid object that the sound "beam" strikes, whether it be a submarine, a surface ship, or a school

SONAR comes from the U.S. Navy designation Sound Navigation And Ranging. A device used for underwater detection, ranging and depth measurement, it operates on the principle of reflected sound waves.

of fish, reflects the waves back to their source, where they are picked up on the sonar equipment. Since the speed of sound in water is known, all the sonar operator has to do is measure the time it took for the waves to go from his own ship to the target and back. He then knows the *range*, or distance, he is from his target. The sonar echo is characterized by a "pinging" noise which the operator picks up on his equipment. Incidentally, a good sonar operator must be very well trained to tell the differences among various water sounds. Otherwise, he may make the mistake of reporting the presence of a submarine to his commanding officer when, in actuality, the "submarine" may turn out to be a whale!

When you hear a cheerleader at a football game shouting through a megaphone, you are also hearing something that depends on the

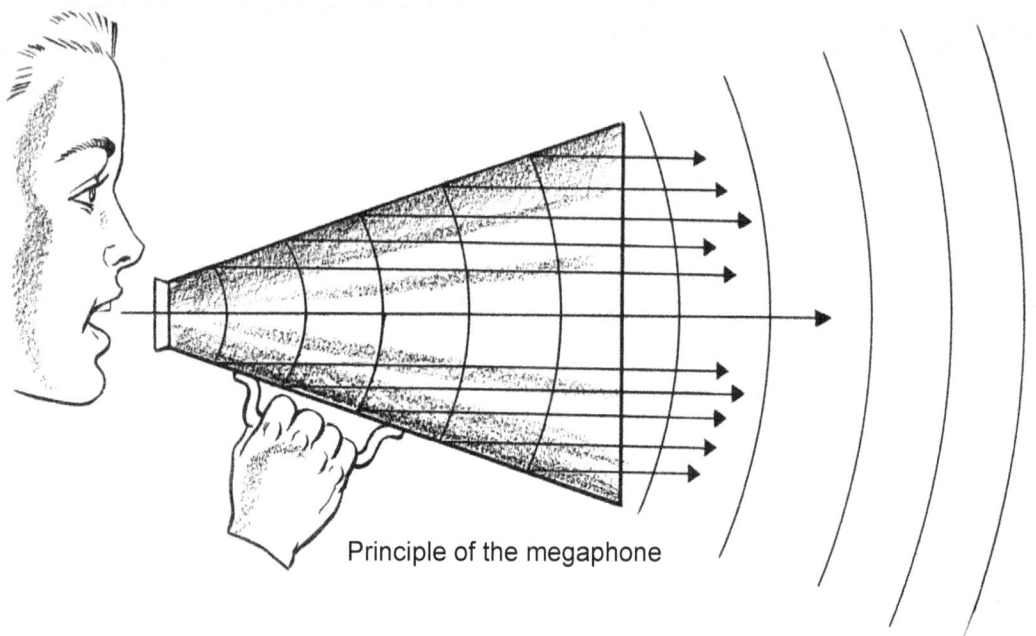

Principle of the megaphone

reflection of sound waves. A megaphone directs sound much as a flashlight directs a beam of light. When the cheerleader's voice is directed into it, the sound waves created by his vocal cords reflect from the sides of the speaker cone and come out concentrated, or "beamed," at the other end. When you cup your ear to hear something better, you are using the megaphone principle in reverse. That is, you are "beaming" sound to yourself instead of to someone else, as you would with a megaphone.

Multiple Echoes

Multiple echoes (more than one echo) can be heard where there are several reflecting surfaces located in favorable positions. When sound waves are able to bounce around between neighboring or nearly parallel surfaces, they sometimes cause the original sound to be repeated again and again. A thunder roll, for example, is heard because of its repeated reflections from cloud to cloud, or from cloud to earth. A bat flying about in pitch darkness is guided entirely by the echoes of its own cries rebounding from various surfaces. Yet these sounds are so high in pitch that most human beings are unable to hear them!

52

There are many examples of multiple sound echoes, as in the canyons of the Rocky Mountains and Arizona. In one old chateau near Milan, Italy, a forty-fold echo can be heard. When a guide in the Baptistry at Pisa sings a few quick notes in succession, the echoes combine and for seconds thereafter a sound like an organ persists. Amazing echoes can also be produced in Mammoth Cave in Kentucky, 300 feet below the earth's surface. Likewise, in certain buildings such as St. Paul's Cathedral in London, the Capitol building at Washington, and the Mormon Tabernacle in Salt Lake City, curved walls or ceilings reflect sound waves from certain points in surprising fashion. Whispers can also be heard many feet away because of reflected sound waves "creeping" along curved surfaces known as "whispering galleries," such as the one in Grand Central Station in New York City.

Making the Whispering Gallery "whisper" in New York City's Grand Central Station.

Acoustics - Reverberation and Interference

THE term *acoustics* (from the Greek word meaning "hearing") is defined as the science of sound. It is concerned with sound's production, its transmission, and its effects. In the more everyday use of the word, however, acoustics is concerned with sound problems that occur in enclosed spaces such as buildings.

Two problems that arise in building acoustics are *reverberation* and *interference*. In designing and constructing buildings, acoustical engineers and architects are constantly concerned with both, especially in auditoriums or broadcast studios where good acoustical quality is important.

Reverberation refers to the repeated reflection of sound waves from smooth surfaces in an enclosed space. In other words, it is a *resounding* or continuing process, like a series of echoes. When you

Reverberation. Sound waves from a TV loudspeaker bouncing back from various surfaces of a room.

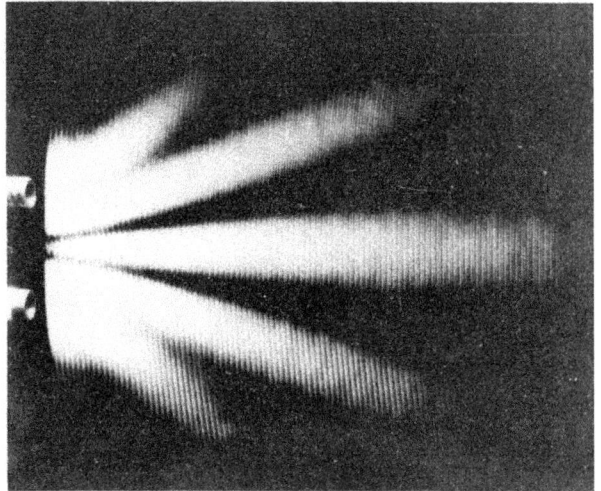

Sound waves made visible by new Bell Labs technique. Left, sound waves spreading out from a telephone receiver in circular condensations and rarefactions. Right, a demonstration of interference. Waves from two sources reinforcing each other at light areas, cancelling each other in dark areas.

are watching TV in your living room, the set's loudspeaker causes sound to reverberate around you. That is, it bounces back from the various surfaces and objects in the room. Unless enough sound is absorbed by your surroundings, each new sound wave would have to compete with the slowly dying reverberations of the sounds that came before. Can you imagine what unpleasant reverberation would strike your ears if you were listening to TV in a completely empty, steel-enclosed room?

Interference is an effect produced by certain combinations of two or more wave trains that reinforce, cancel, or otherwise interfere with each other. When, for example, two sound waves occur at the same time and are *in the same phase*—that is, when their condensations and rarefactions coincide with each other—the waves strengthen or *reinforce* each other; in other words, they are louder. But when the rarefactions of one coincide with the condensations

Johns-Manville

Sanacoustic Units being installed in a ceiling. Hundreds of tiny sound-absorbing holes have been drilled in these panels to trap and deaden sound waves.

A modern, completely soundproofed broadcast studio. Actually, this is a room-within-a-room, specifically engineered for maximum sound isolation under broadcast conditions.

Johns-Manville

of the other (that is, when they are *out of phase*), they cancel each other and silence results.

Acoustical engineers know that the human ear needs about 1/15th of a second between two sounds if it is to recognize that there *are* two sounds instead of just one. This works out to mean that unless you are at least 37 feet from an echoing surface you will hear only one *prolonged* sound instead of separate ones.

Small rooms reverberate by reflecting sound from wall to wall so that syllables of words are each "dragged out" by other reflections joining with them. But in larger rooms like auditoriums, separate echoes coming from faraway sources sometimes arrive at the ear in time to cause confusion with *direct* sounds that come, for instance, from a speaker on a stage.

When such troublesome sound reflections are present, they may be reduced by muffling or *soundproofing*. In recent years, a whole new industry has been based on the deadening or *absorbing* of offensive sound waves by means of special wall materials, soft draperies, floor coverings, asbestos tiling, "mineral wool," and the like. Schools, hospitals, and assembly rooms, especially, use such materials in their ceilings and walls to trap and deaden sound.

In some theaters and lecture halls, special sound reflectors are used to direct the sound toward listeners and to prevent it from reaching flat echoing surfaces. Have you ever noticed that you can hear speech more clearly in an assembly room filled with many people? This is because the audience (together with the clothing they are wearing) breaks up the echoes that interfere with distinct hearing. The same is true with a soft blanket of snow covering the ground in winter. Like an acoustical tile, for example, the snow contains a great number of small holes between its crystals. When sound waves enter these tiny holes, they cannot get out.

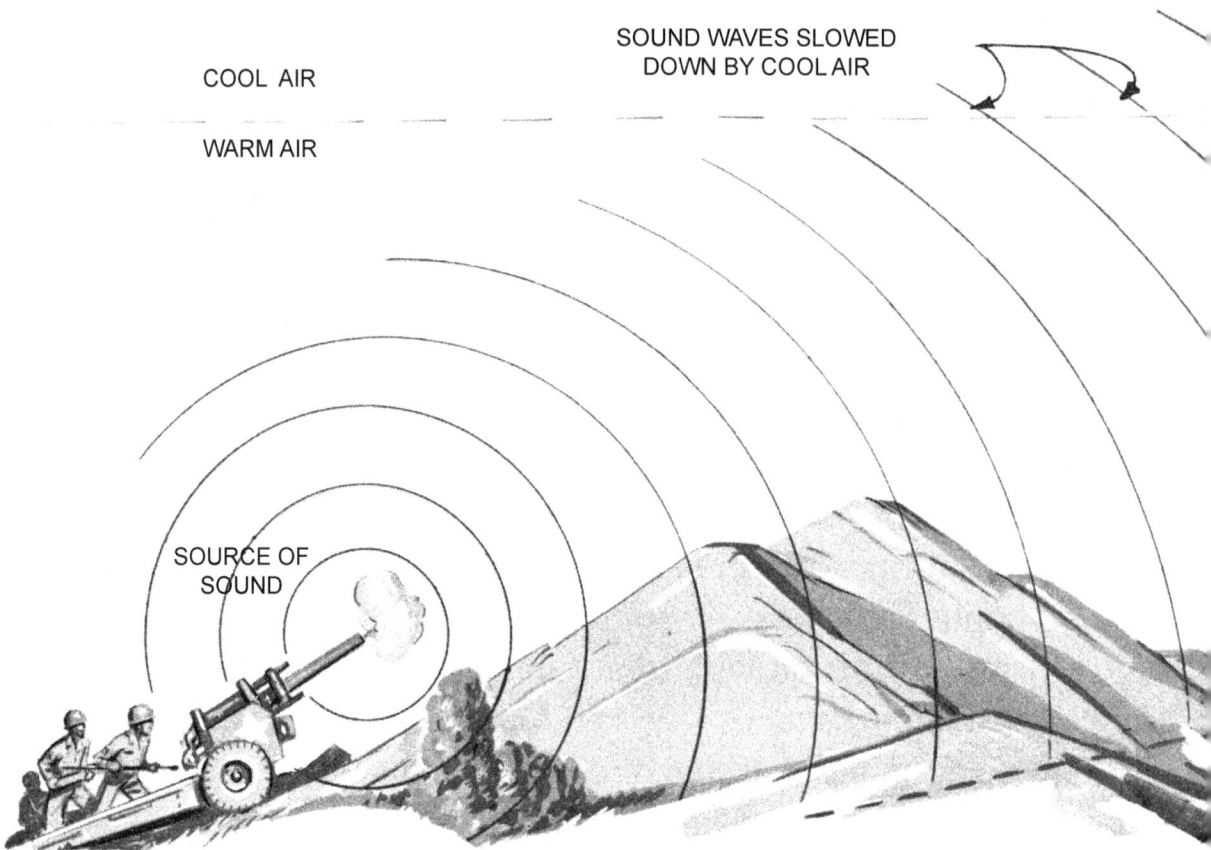

COOL AIR

WARM AIR

SOUND WAVES SLOWED
DOWN BY COOL AIR

SOURCE OF
SOUND

"Zones of Silence"—How Sound is Refracted

WHEN light rays pass from one medium into another, as from air into a piece of glass, they are bent, or *refracted*. The same thing is true of sound waves.

Let us say that a squad of artillerymen fires a gun during the daytime. The sound wave created by the explosion travels away from the gun's muzzle, expanding uniformly like a huge globe in all directions above the surface of the ground. Eventually the upper part of the wave front reaches a region of air that is cooler. It is slowed down, because sound travels slower in cool air; and, during the daytime, air is warmer nearest the ground, cooler farther up. Meanwhile, the lower part of the wave front, being close to the ground

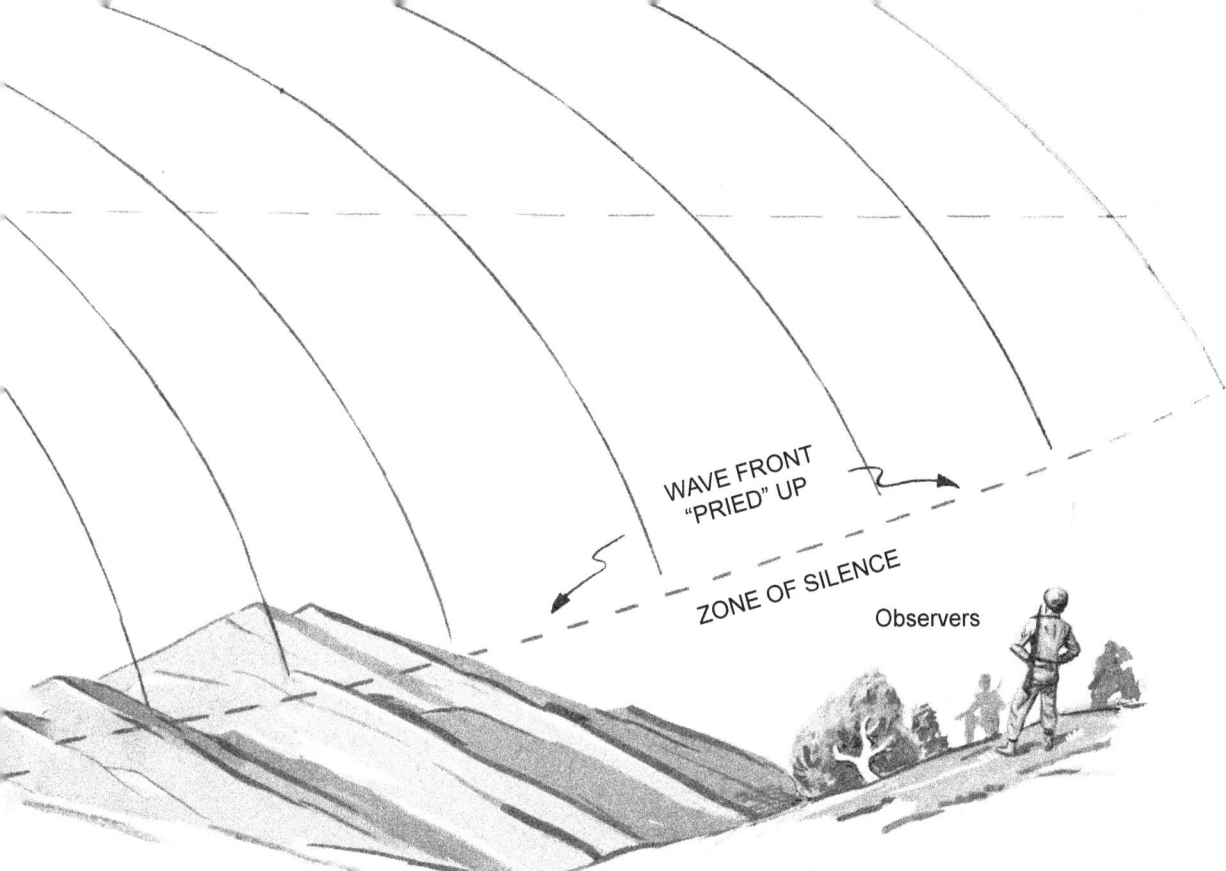

WAVE FRONT "PRIED" UP

ZONE OF SILENCE

Observers

in warmer air, shoots ahead as before. Thus the upper and lower tips of the wave front are traveling at different speeds. The result is that the entire wave front, including the rest of the waves that follow it, are "pried" upward, creating a "zone of silence" underneath. Observers located in such a zone, even if it is only a few miles away from where the gun was fired, will never hear the explosion.

Some historians say that Napoleon's defeat at Waterloo was partly due to "zones of silence." Troops which the French commander expected to move up on an agreed signal of cannon shots simply did not hear them because they happened to be located in one of these silent areas.

During the night, particularly near the water where the temperature tends to be warmer, the opposite can happen. Sound waves that would ordinarily be refracted upward are instead directed back toward the ground. As a result, our ears can often pick up faraway sounds at night that we would not hear during the day.

During the daytime, however, air that is as much as thirty or forty miles above the earth tends to be a bit warmer than the air beneath it. So it may be that sound waves that are originally refracted upward are then refracted back toward the earth again, where observers approximately one hundred miles away hear them. Thus it might possibly have happened that Napoleon's cannon signal at Waterloo actually reached the ears of his English enemies instead of his own troops.

What is Resonance?

SUPPOSE you are in a playground pushing a child on a swing. Each time, just as the swing starts to move away from you, you give it a

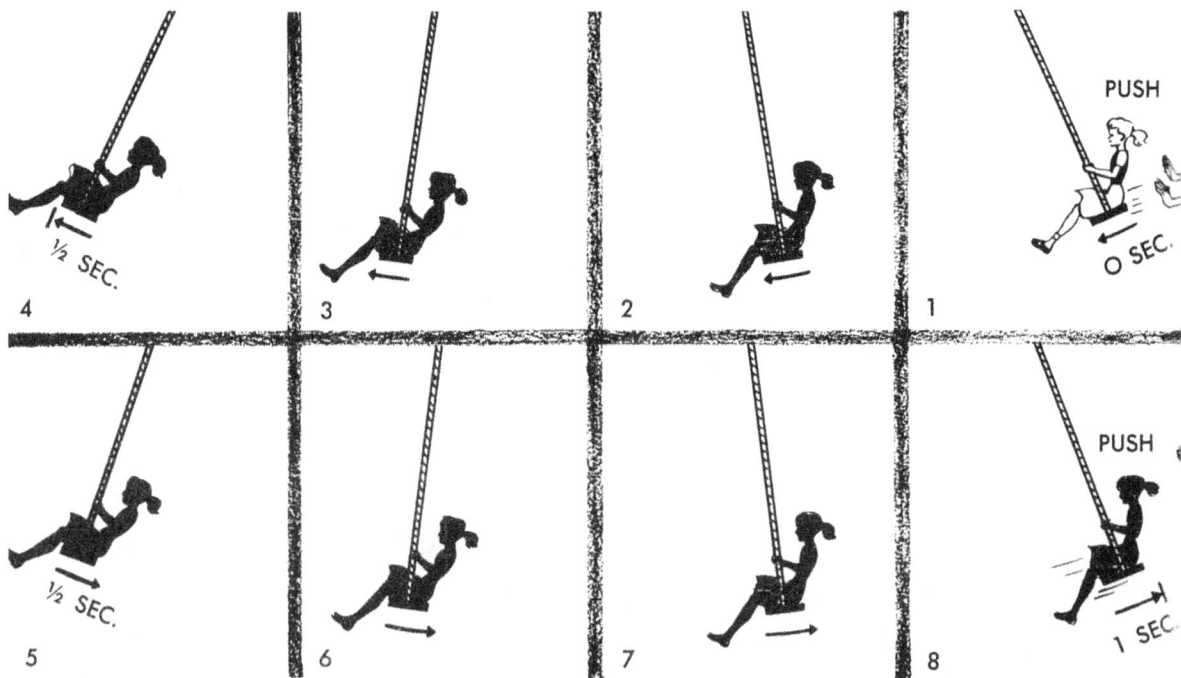

PUSH

½ SEC.

O SEC.

4

3

2

1

½ SEC.

PUSH

1 SEC.

5

6

7

8

little push. If the swing moves back and forth in an arc once a second, you could say that its frequency of free vibration is one *cycle* a second; and, you yourself will give exactly one push a second. If you gave more or fewer pushes than that, you might stop the swing's action altogether.

In other words, your pushes exactly agree with the frequency of the swing's vibration. Or we might say that your pushes are *in resonance* with the swing's vibration.

An object *resonates* when it is caused to vibrate by another sounding body whose frequency of free vibration is exactly or very nearly the same as its own.

Sometimes resonance is called *sympathetic vibration*. You have probably heard of delicate wine glasses being shattered when certain notes are sounded on a musical instrument. The glass first resonates in "sympathy" with the instrument's notes as their sound waves strike it, vibrating at its own free-vibration frequency. Then, when this vibration-rate becomes so rapid that the amplitudes of the glass molecules cannot "take it," the glass breaks.

61

Likewise, there is the famous story of a company of soldiers who once crossed a certain flimsy bridge. The soldiers marched along evenly in parade step, and the bridge swayed more and more until it finally broke. The reason it did so was that the frequency of the soldier's steps was unfortunately just right to set the bridge into resonance, producing larger and larger sways until the structure collapsed.

Blowing across the top of a bottle will also produce, by resonance, a note by the air enclosed in it. Lord Rayleigh, a British scientist who contributed much to the study of sound, found that blind persons could judge the size of a room by its resonance to stray noises.

You can easily demonstrate resonance by getting a piano to "answer you." Step on the pedal that lifts the felt hammers off the strings, then sing a quick loud note into the piano. Put your head down to the piano and listen; you will hear those strings resonating that match the frequencies of the note you just sang.

When you hold a seashell up to your ear and hear a roaring noise, you are not hearing the "sea," but simply all the stray noises about you, made greater by resonance of the air column in the shell.

One of the best examples of resonance occurs in the human voice. The vocal cords are tough pieces of muscle, and they vibrate when air is passed through them. By changing the position of our jaw, tongue, and cheeks, we change the shape of our mouth cavities so that they respond, by resonance, to various overtones produced by the vocal cords' vibrations. Thus we can change the quality and, in part, the pitch of our voices.

Ultrasonics—the Mysterious World of "Silent Sound"

AS FAR as man is concerned, any vibrations below 16 times per second or above 20,000 times per second do not cause sound in his ear. For this reason any such vibrations below or above these limits are sometimes called "silent sounds."

The science of *ultrasonics* deals mainly with vibrations (frequencies) that are above the 20,000 cycle range.

Ultrasonic waves, being of short wave length, have the characteristics of all short waves; namely, they have high energy and they travel in straight lines. They are put to astonishing uses in the fields of medicine, industry, and biology. Sonar, which we discussed earlier, uses these ultrasonic waves.

A special oscillator, called a piezo-electric crystal oscillator, is used in scientific work to produce ultrasonic waves. They can also be produced by the rapid changes in a length of iron bar that is magnetized by a high-frequency alternating current. By using these devices, frequencies up to 12 million vibrations a second have been produced.

As we mentioned earlier, a dog's ear has the ability to hear sounds of much higher frequency than the human ear. By means of an ultrasonic whistle, dogs used in police or military work can receive "silent orders."

In laboratory work, high-frequency ultrasonic vibrations can be

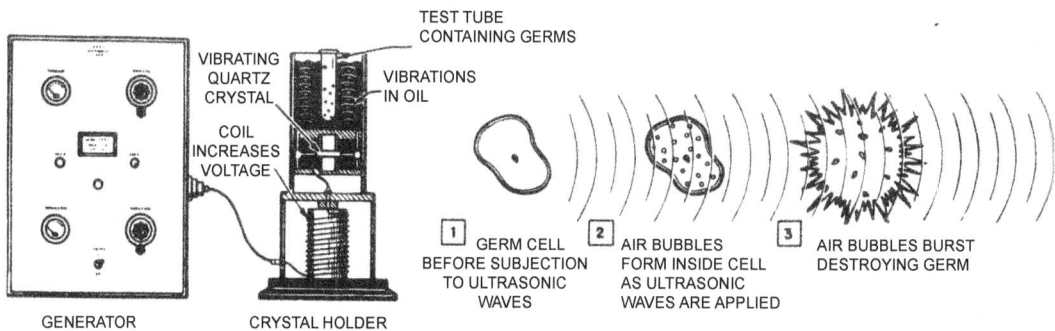

How ultrasonic waves can kill germs

produced by a special device known as a *transducer*. It is controlled by an electronic oscillator, and operates submerged in a chemical liquid. By an adjustment of the oscillator's frequency, ultrasonic waves can be sent through the liquid. They have a number of practical uses. When dirty metallic objects are placed in the liquid they are rapidly cleaned. Virus and bacterial particles can be torn apart by such waves, and can thereby provide scientists with an invaluable aid in their studies of the nature of life.

An "ultrrasonic bath." Electric shaver heads being cleaned by action of sound waves directed through a solvent. Vibrating quartz crystal underneath the trough issues the ultrasonic waves.

An "ultrasonic fountain." Dime-sized metal disk suspended in tank issues sound waves at 3,000,000 cycles per sound when voltage is applied across it. Slightly hollow, the disk focuses the waves upward causing the water jet.

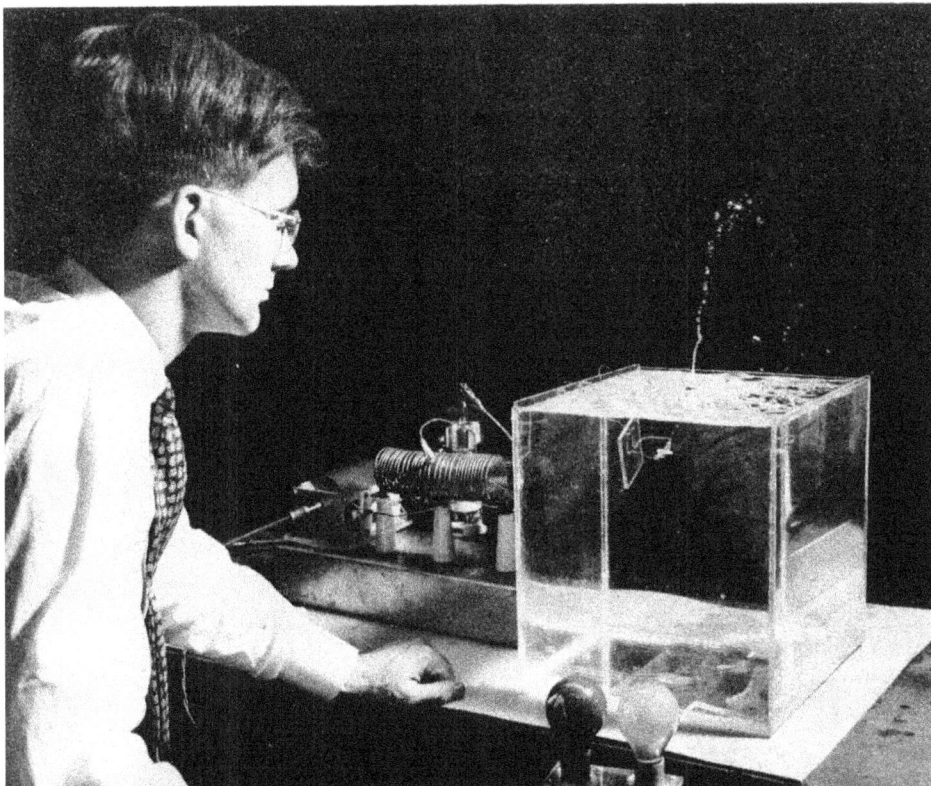

Getting an ultrasonic OK. Crankshafts being tested for flaws by a *relfectoscope*. Ultrasonic vibrations are beamed through the part tested, reflected back, and amplified (increased in size) on a cathode tube screen. Time elapsed of arrival of vibration "pips" on screen indicate flaws to the operator.

Testing large rotor forgings with the reflectoscope. Automatic signaling devices indicate the exact location of defects for quick corrective action.

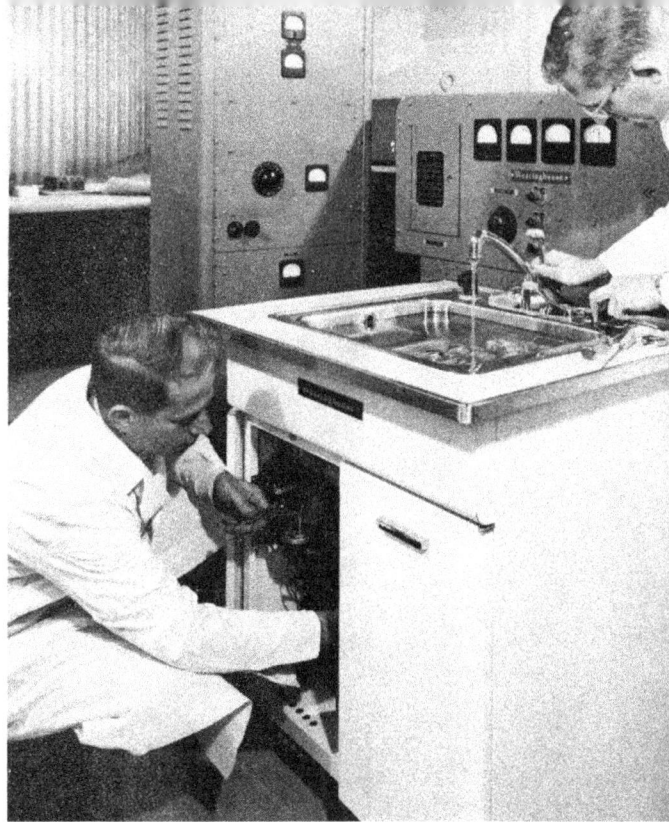

Westinghouse Electric Corporation

Cleaning with sound. Housewives of the future may be doing their dishes and other cleaning chores with an ultrasonic sink, such as the experimental Westinghouse model shown here. It has a new and improved transducer.

By exposure to ultrasonic waves, foods can be quickly and thoroughly sterilized without heating them. If seeds are placed in the path of such waves, more of them will sprout. By observing the way in which ultrasonic waves are reflected through metal castings several feet thick, engineers can quickly check them for flaws.

We are only beginning to realize what tremendous energy ultrasonic waves have. Frogs, mice, and other small animals, when placed in the path of ultrasonic waves, have been deafened or killed. Milk can be pasteurized or homogenized, and an egg can even be cooked *without breaking its shell*, by exposure to ultrasonic waves.

Ultrasonics has also been known to clear away fog and smoke. Cities that have smog may one day use ultrasonics to rid themselves of it. In medicine, high-frequency ultrasonic waves have even been used to perform brain operations *without opening the skull*.

It is certain that modern research in ultrasonics will reveal to the world of tomorrow a wide variety of useful applications.

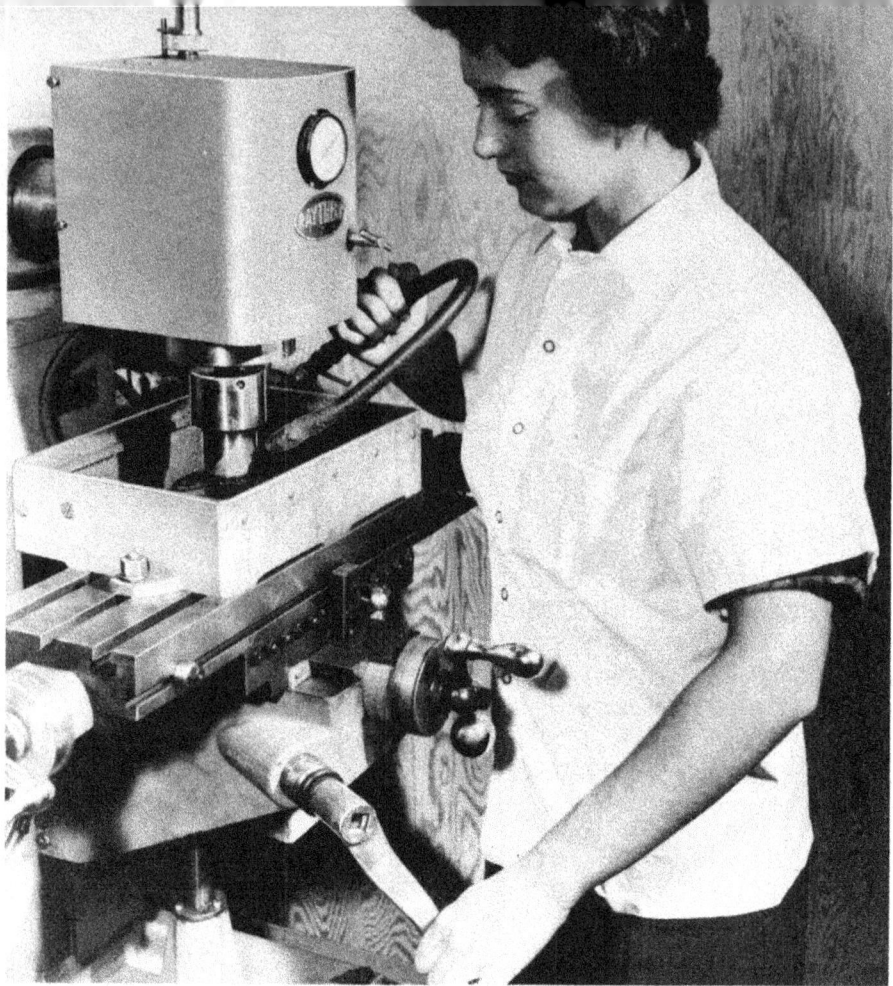

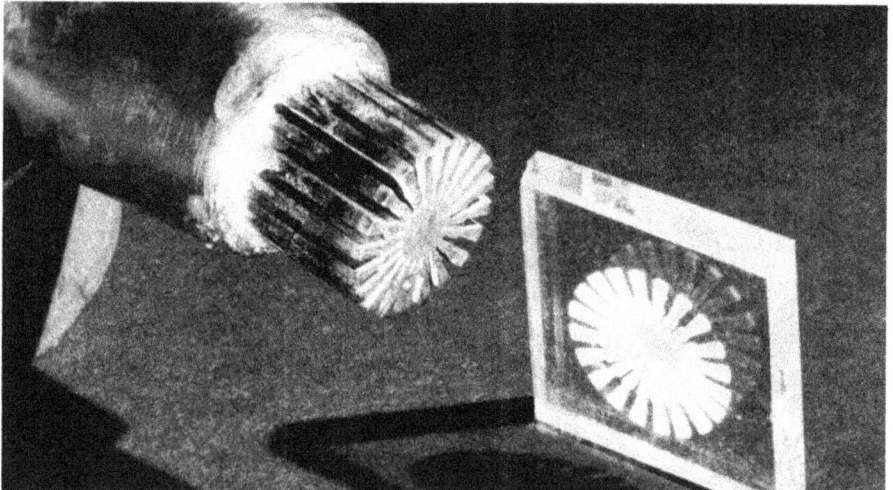

This Raytheon ultrasonic impact grinder can slice neatly through hard materials such as quartz, glass, and precious stones. The cutting tool does not revolve; instead it vibrates up and down 25,000 times a second. Meanwhile, a liquid abrasive flows between it and the work. Particles of the abrasive are driven with great force against the work, thus cutting out an exact copy of the tool's shape.

Supersonics

THE term *supersonics* should not be confused with the term *ultrasonics*. While ultrasonics deals with frequencies above a certain range, *supersonics* deals with the *motion* of objects that move at speeds higher than the speed of sound. And, since it would be difficult indeed to get anything to travel faster than sound itself through any medium other than air, the problems of supersonics are pretty much limited to planes and missiles that fly through the air.

When an object is traveling at a speed which is less than that of sound, its motion is called *subsonic*. In air, the disturbance created by such an object runs far ahead of it. This is why Londoners, for example, during the Battle of Britain in World War II, could hear the whistle of bombs long before they struck their targets. Subsonic moving objects also leave a wake behind them, much like the wake of a ship.

The situation is different, however, in the case of an object tearing through the air at supersonic speed. This time, no disturbance runs ahead of the object because the object is going faster than the speed of its own sound waves. This motion creates an interrupted air pressure wave around the leading edge of the object; this is known as a *shock wave*. The air surrounding the moving supersonic object simply does not have the time to adjust itself to this interrupted pressure.

Because these interrupted pressure areas move right along with an object traveling at a supersonic speed, scientists call them *standing shocks*. It is because of these standing shocks that supersonic aircraft undergo such great stresses and vibrations; in fact, many experimental planes have been torn apart by them. Great care and planning therefore must go into designing aircraft for ultra-high

NACA Photo

A model missle streaking along at 2,500 miles per hour in supersonic wind tunnel of National Advisory Committee for Aeronautics' Ames Laboratory, Moffett Field, CA. The shadowgraph photo shows shock lines streaming back from the model's needle nose and tail.

This is what 10,000 miles per hour look like! The metal sphere in this shadowgraph photo has just fired from a new light-gas gun developed by NACA scientists. Made at one of ten millionth of a second, the photo shows the strong shock waves formed by the speeding model. The knowledge gained in such experiments is of great value in designing ICBM's.

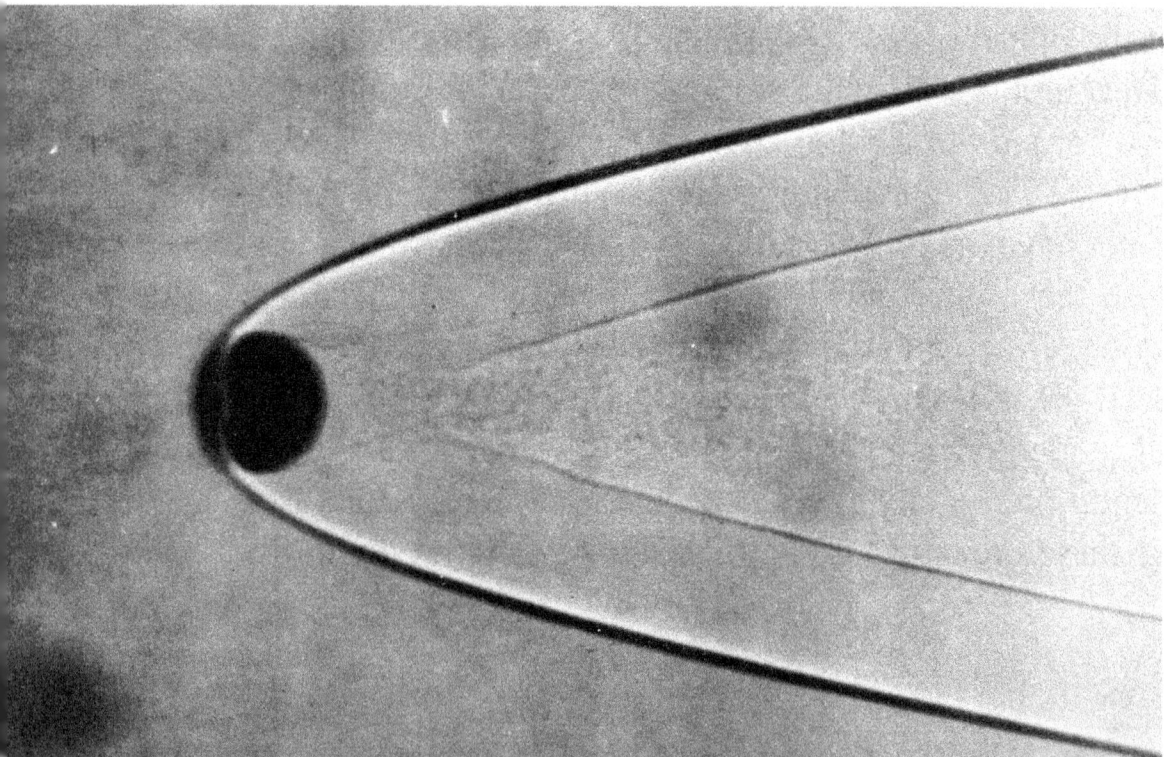

NACA Photo

speeds. For example, swept-back wings, very pointed noses, and other streamlining features have been found to lessen a supersonic plane's resistance to air.

Formerly, it was thought that an airplane could go no faster that the speed of sound. Today, however, specially trained and highly skilled pilots have many times gone beyond it.

What happens when one of them does so?

Breaking the Sonic Barrier

WHEN a guided missile or an airplane travels through the air at a subsonic speed, it meets the resistance of the air. This air resistance is called *drag*. As the plane or missile *increases* velocity, the drag *lessens*. It continues to do so until the plane or missile approaches the speed of sound. When the speed of sound is reached, the plane or missile requires a lot more energy to overcome the standing shocks. Thus the drag increases, this time very sharply.

This sharply increased air resistance, met in going from subsonic to supersonic velocity, is called the *sonic barrier*. When the plane or missile "breaks through" the sonic barrier, a loud noise is heard by observers on the ground; it is sometimes called the *sonic boom*. This explosive sound is caused by the sudden breaking up and scattering of intense pressure areas around the aircraft. The pilot of the plane does not hear the sonic boom because his own speed outruns the speed of sound. After the missile or plane goes faster than the velocity of sound, however, the drag increase slows down again to normal.

Scientists and airmen use a special term to indicate the ratio between the speed of a flying object and the speed of sound in the surrounding air. It is called a *Mach number*, named for Ernst Mach

Supersonic wind tunnel, used to test full-size airplanes at flying speed and to investigate high-speed aircraft in the landing-speed range.

Sharply swept-back wings of this test model are generally favorable for supersonic flight. The wings on this model have porous edges to give them better performance at lower speeds than that of sound.

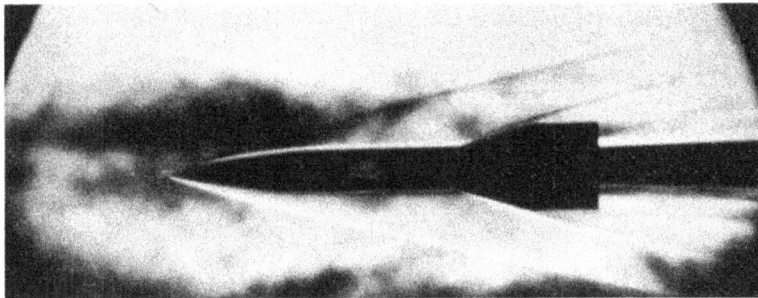

A high-speed camera records the development of shock waves around an experimental model in a supersonic wind tunnel. Airspeed in the upper frame was at a Mach number of 2.8. In the middle frame, the Mach number was 4.6. In the lower frame, the Mach number was 6.3. At sea-level temperatures, these speeds are equal to about 2,100, 3,500, and 4,900 miles per hour, respectively.

(pronounced Mahk), an Austrian physicist. Mach number 1 has been set as the speed of sound. If a plane is traveling at Mach number 2, it is traveling at twice the speed of sound. For subsonic speeds the Mach number is less than 1; for supersonic speeds it is greater than 1. A shock wave that is set up by an object traveling with a Mach number above 1 is known as a *Mach wave*.

73

SHOCK WAVES AT EARLY
STAGE OF EXPLOSION

GASES
PRODUCED BY
ATOMIC
EXPLOSION

EXPANDING GASES
PRODUCED BY
ATOMIC EXPLOSION

SHOCK WAVE SEPARATING
AT LATER STAGE

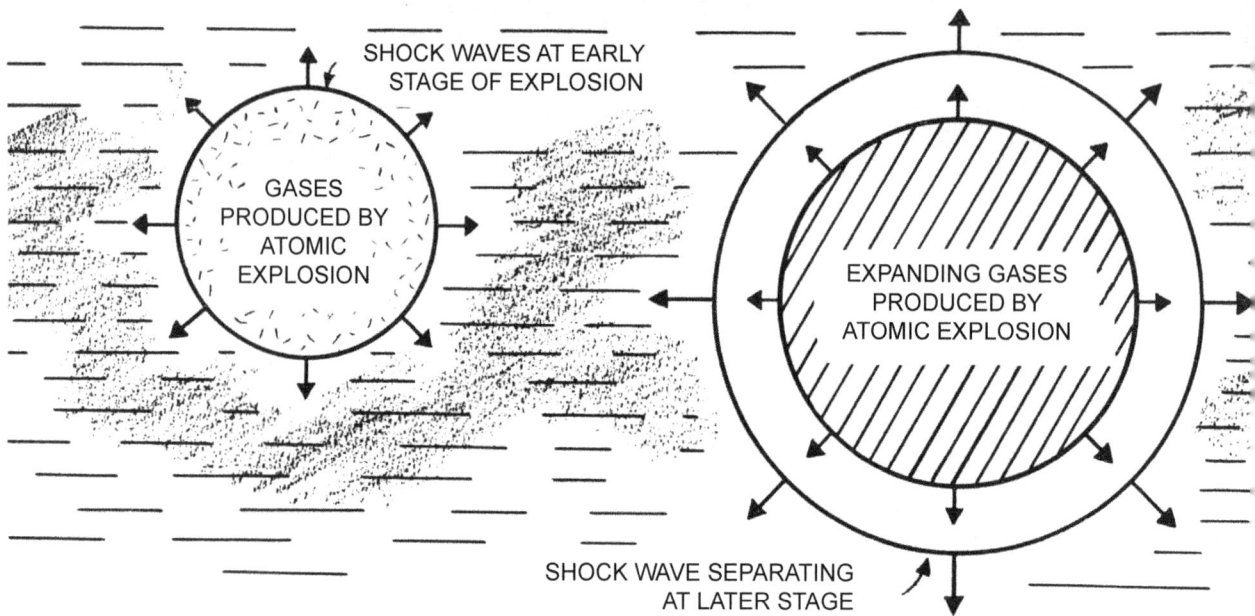

How shock waves behave at early (left) and later stages of an atomic explosion.

Sound and the Atomic Bomb

SHOCK WAVES are also produced in various kinds of explosions. When, for example, an atomic bomb goes off, it causes material to move outward at very high supersonic speeds. But, whereas the supersonic airplane or guided missile advances with a constant speed (driven by its motors), the hot gases released by the bomb lose their power and slow down as they expand in volume. When this happens, the shock wave separates from the hot gases and starts hurtling through the surrounding medium "on its own."

It is not necessarily the exploding matter itself, but the shock wave that causes such terrific damage in an atomic or hydrogen bomb explosion. Such a wave can flatten buildings as if they were made of matchsticks. During the H-bomb testing at Eniwetok Island in the Pacific, it was a shock wave that turned a whole coral island into a pool of water about one mile wide! And it is the underwater shock wave produced by depth charges that damages or destroys submarines.

"Operation Crossroad" - code name for the Able Day atom burst during tests at Bikini Atoll. This excellent photo shows the shovck wave sweeping out over target ships toward the beach. Note that it is already outrunning the explosion gases.

How Sound is Reproduced

SOUND WAVES can be reproduced in many ways. They may either be reproduced so that we can hear them right away, or they may be preserved so that we can hear them later by some mechanical method.

Sounds are reproduced for immediate hearing by telephone, radio, or TV systems. The telephone reproduces and receives sounds from the human voice transmitted from a distance over wires, by means of electricity. Radio and TV systems do the same thing by electrical means, but without a connecting wire.

Sounds for later reproduction (or "playing") are stored by phonograph records, various recording systems, and in the sound tracks of films.

In phonograph recording, sound waves are received by a flexible diaphragm (a thin, vibrating plate) to which is attached a needle. The waves cause the diaphragm to vibrate so that the attached needle scratches grooves in the soft wax of a disk. The grooves, of course, vary with the frequency, amplitude, and complexity of the sound waves. When these grooves are retraced by another need attached to a "playing" diaphragm, the vibrations of the original sound waves will be heard with considerable fidelity. In office dictating machines, a cylinder is used instead of a disk.

More accurate reproduction of sounds is made by electrical recording. Sound waves are changed into electrical impulses by means of a microphone and amplifier system. These impulses in turn drive a recording needle by means of an electromagnet. When the recording is played back, another electromagnetic devise is used to send the electric impulses through an amplifier and into the loudspeaker of a radio receiver.

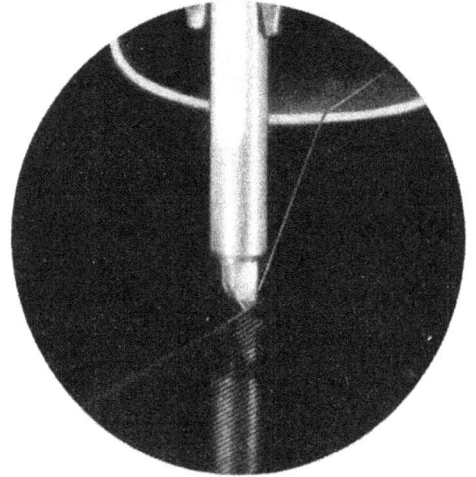

A sapphire cutting needle, or *stylus* (lower right), scratching grooves on a record. The to-and-fro motion of the stylus displaces the grooves in proportion to the sound it receives, making "wiggles" that can be seen on any record under a microsope (upper right). At left, the sound track on a motion picture film.

In motion picture films, a *sound track* runs parallel to the pictures on the film. When the film is "shot" at the studio or "set," the sound track is made by changing the sound variations into variations of light, which are then photographed right on the film. When the picture is shown at the theater, the light variations are then changed back into sound.

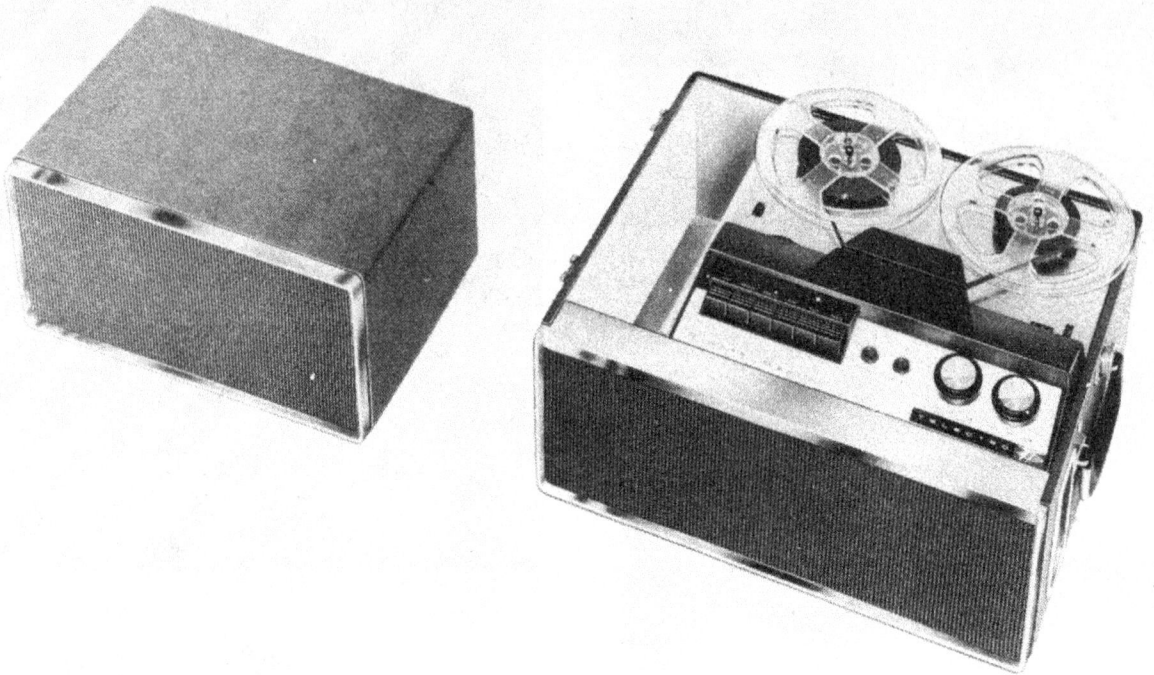

Telectrosonic Corporation

A modern portable tape recorder with sterophonic record-and-playback features.

In magnetic recording devices, the sound waves are changed into electrical currents. These currents alter the magnetic structure of a metallic ribbon. To reproduce the original sound waves, the metallic changes are turned back into electrical currents which then operate a loudspeaker.

In any of these methods of sound reproduction, the more accurately the device responds to changes in frequency, amplitude, and overtone combination—and the more accurately it plays them back—the more satisfactory the reproduction will be.

Sound and You

SOUND is everywhere about you every minute of the day. Your ears are literally bathed in it from morning till night. You can do more than simply "hear" sound; you can train yourself to *picture* it as well.

Begin right now by listening for the very next sound that reaches your ear. What kind of *vibration* probably caused it? Was the *pitch* high or low? The *quality* pleasant or unpleasant? Did you hear an *echo*? Picture the *condensations* and *rarefactions* pulsing out in ever-widening circular waves through the air molecules in all direction. Imagine the tiny divisions of them that enter the *canal* of your ear.

Try your best to understand sound. It will be around as long as you are.

A new idea in supersonic aircraft design. The delta wing Convair F-102A has been given a "wasp waist" so that it will slip more smoothly past the sonic barrier.

Convair

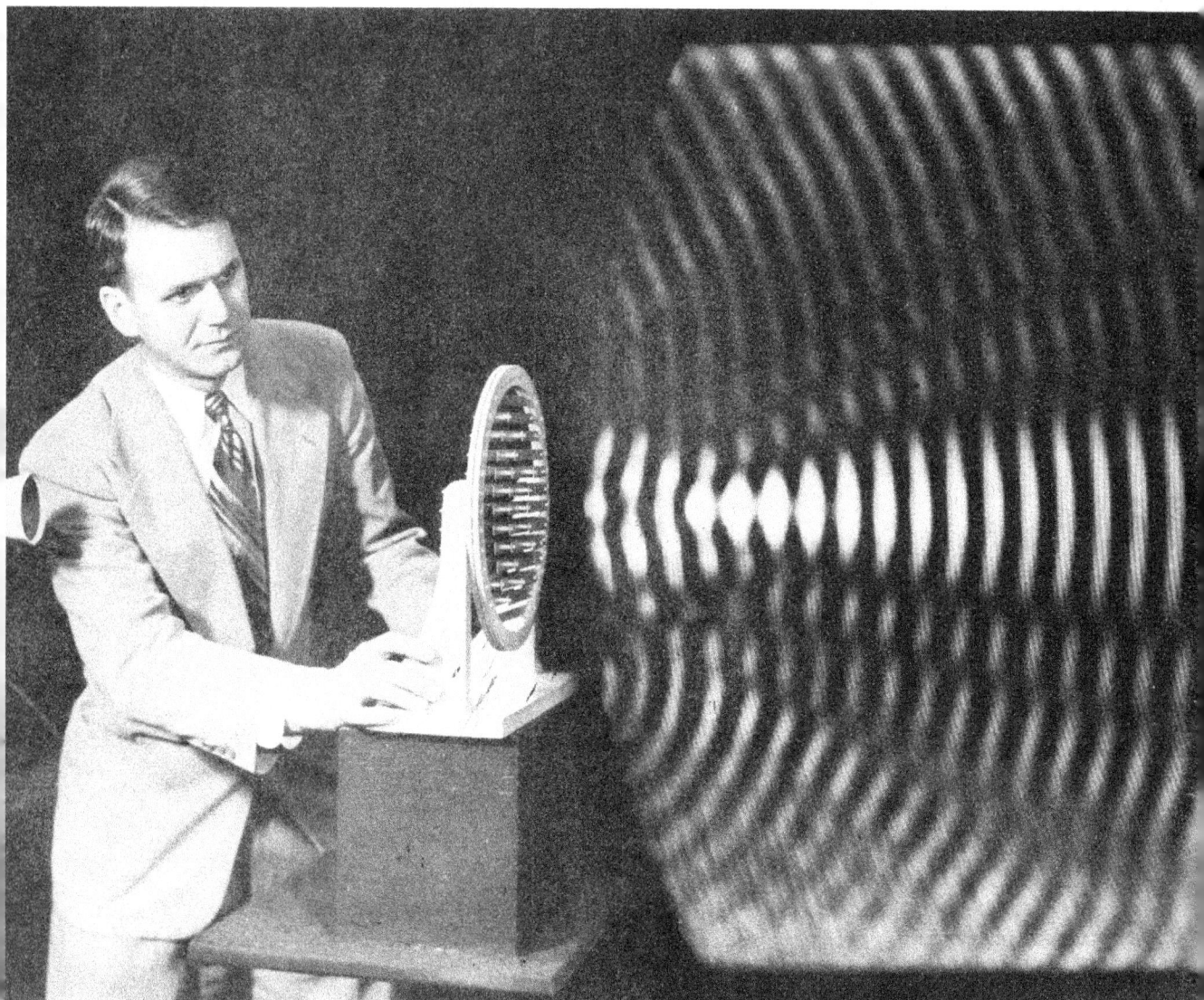

A visible pattern of sound waves. This new technique of studying sound demonstrates the focusing effect of an acoustic lens on sound waves issuing from the horn at the extreme left. The wave pattern is produced by a scanning technique, somewhat like that used in television.

A CHECK-LIST OF SOUND FACTS

You can quickly refresh your memory on the important points in this book by reading the condensed summary below.

• Sound waves are produced by vibrating bodies.

• Sound needs a material medium through which to travel, or there is no sound.

• Sound cannot travel in a vacuum.

• Sound is transmitted in all directions from its source.

• Sound waves can be transmitted by solids, liquids, and gases.

• Air is the commonest medium through which sound comes to us.

• The speed of sound in air at 0° C. (centigrade) is approximately 1,100 feet per second.

• Sound travels faster in water than in air, and faster in most solids than in water.

• A sounding body must vibrate at least 16 times a second to be heard.

• Most human ears cannot hear vibrations that are much faster than 20,000 times a second.

• The number of vibrations a sounding body makes per second is known as the **frequency** of that body.

• Everything in nature vibrates and has its own natural rate of free vibration.

• The fact that air is so easily set in motion makes hearing possible.

• To hear sound, three things must be present: a vibrating body to produce sound waves, a medium to carry the sound waves, and the human ear system to receive them.

• The sound of the human voice is produced by the vibration of the vocal cords in the larynx when air is forced through the glottis.

• Sound waves are longitudinal (back-and-forth) waves consisting of **condensations** (crowded up places in the medium) and **rarefactions** (stretched out places in the medium).

• The **wave length** of a sound wave is the distance from one condensation to the next; the **amplitude** is the distance that a single particle, as an air molecule, is displaced during the wave's passage.

• The human ear can hear wave lengths ranging in length from about one inch to seventy feet.

• The greater the elasticity of a medium, the better and faster it will transmit condensations and rarefactions.

• As the density of a medium (the amount of it per given unit) increases, the speed of sound grows less.

• Sound waves travel faster in air as the air gets warmer.

• Sound waves travel 2 feet per second faster with each centigrade degree rise above 0°.

• At any given temperature, ordinary sound waves travel with the same speed through the same medium.

• **Pitch** refers to highness or lowness of sound and is determined by a sounding body's frequency of vibration. The greater the frequency, the higher the pitch; the lower the frequency, the lower the pitch.

• **Loudness**, or intensity of sound, is determined by the force with which sound waves strike the ear.

• The unit expressing changes in intensity of sound is the **decibel**.

• **Quality** distinguishes one musical instrument from another and one human voice from another when both have the same pitch and loudness; it is determined by the number and intensity of the overtones produced by vibrations in divisions of the sounding body.

• The **fundamental** is the lowest tone produced by a sounding body when it vibrates as a whole.

• Sounding bodies can vibrate at several different frequencies at the same time; sections of such bodies vibrating above the fundamental are called **overtones.**

• No two musical instruments, as well as no two human voices, ever have the exact same quality.

• Musical instruments are divided into three groups: **wind, stringed,** and **percussion.**

• An **echo** is the repetition of a sound, caused by reflected sound waves.

• **Sonar** makes use of reflected high-frequency sound waves to detect the presence of objects under water.

• **Multiple echoes** occur when there are several reflecting surfaces located in favorable positions

• **Acoustics** is concerned with sound's reproduction, its transmission, and its effects. In physics, it is the science of sound.

• **Reverberation** refers to the repeated reflection of sound waves from smooth surfaces in an enclosed space; it is a resounding or re-echoing process.

• **Interference** occurs when two sound waves unite; if their condensations coincide, louder sound results; if condensations and rarefactions coincide, silence results.

• The industry of soundproofing is concerned with absorbing or deadening of unwanted sound waves in buildings such as schools, auditoriums, and hospitals.

• Sound waves, like light waves, are bent or refracted when they pass from one medium to another.

• An object resonates when it is caused to vibrate by means of another sounding body which coincides with the original object's own frequency of free vibration.

• **Ultrasonics** deals with frequencies that are above the hearable or 20,000-cycle range.

• **Supersonics** deals with the motion of objects which move at speeds higher than that of sound.

• An object moving at a speed less than that of sound is called subsonic.

• The interrupted pressure-wave around the leading edge of a supersonic vehicle is known as a **shock wave.** Since it moves right along with the object, scientists call it a **standing shock.**

• The sharply increased air resistance met in going from subsonic speed to supersonic speed is known as the **sonic barrier.**

• **Mach number** is the term indicating the ratio between the speed of a flying object, and the speed of sound in the surrounding air. Mach number 1 is set as the speed of sound itself.

• Sounds can be reproduced for immediate hearing by telephone, radio or TV. They can also be reproduced and preserved for hearing later by devices like the phonograph, sound tracks of films, and electric and magnetic recorders.

Sound Experiments You Can Do

Besides the many sound experiments in the text of this book, here are some more you can try. For better understanding, reread the text discussions before you perform the experiments.

Find a cone-shaped wax carton such as some soft drinks or milk come in. Cut off the bottom to make a megaphone. Then get an empty coffee can and stretch some rubber from a balloon tightly over the top of it. Sprinkle some salt or sugar grains on top of the rubber surface. Point your megaphone at the can and say "Ahhhhh" into it, letting your voice rise and fall slowly. At certain pitches you will see the salt or sugar particles bounce up and down on the rubber sheet. The rubber is vibrating *in resonance* with the sound waves made by the megaphone.

Get two very long cardboard tubes, an alarm clock, and a piece of board about a foot square. Place the tubes at an angle to each other so that the open ends are an inch or two away from each other, but not touching. Place the board an inch or two in front of the ends of the tubes. Hold the clock at the opposite end of one of the tubes and tell a friend to hold his ear close to the other end of the second tube. The ticks of the clock will be heard, even though the tubes are not connected. The ticks travel through one tube, bounce against the board and are *reflected* into the other tube where they are carried to your friend's ear.

Make a set of "musical flower pots." Select about a dozen pots of different sizes. Suspend them in a row on a long stick with strings. Have each string going through the hole in the bottom of the pot and tied to a small stick inside the pot. Tap each of the pots with a wooden mallet. The small ones will give out notes of *high pitch*; the bigger ones, notes of *lower pitch*. With practice you can play a tune on this *percussion instrument*.

Make a *wind instrument* by taking a rack of test tubes and filling them with different amounts of water. Fill the test tube at one end with just a little water, the next one with a little more, and so on. Blow across the mouths of the test tubes. The test tubes with only a little water make low notes. As the air columns get shorter in the tubes with more water in them, the *pitch* of the notes will rise.

Prove that water is a better conductor of sound than air. Get a watch and a large metal teapot filled with water. Hold the watch tightly against one side of the teapot and place your ear against the opposite side. Listen to the ticking of the watch. Remove the teapot, holding the watch at the same distance from your ear. The ticking of the watch will be much fainter—if it can be heard at all.

Experiment with reflecting sound waves by making a "sound mirror." Find an old radiant type electric heater with a curved reflector. Remove the heating coil and hang a watch at the center on a string. Make an ear trumpet from a roll of newspaper and listen to the ticking. By finding just the right position for the watch the ticking will be very loud.

Make your own home-made siren from an old pie tin. Punch evenly-spaced circular rows of holes in the tin and mount it on an electric hand drill. Start the drill spinning, then blow air through a tube on the various rows of holes. Listen to the sounds of varying *pitch*.

Find out how far away in miles a flash of lightning has struck. When you see the flash, start counting in seconds by saying "one-hundred and one, one-hundred and two," and so on. Stop counting when you hear the thunder-clap. Since sound travels about one mile in five seconds, divide the number of seconds between the flash and the thunder-clap by *five*. The answer will be the distance in miles from where you are standing to the place where the lightning struck. For example, say you count off just ten seconds before you hear the thunder. Divide ten by five. The flash was two miles away from you.

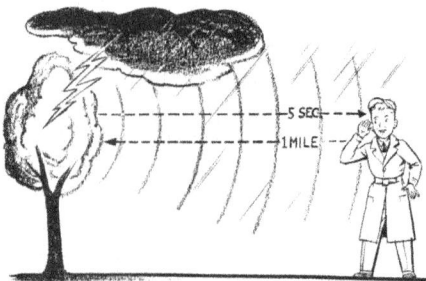

Demonstrate the *Doppler effect* for yourself. Take a length of rubber hose about five feet long and fit a toy whistle securely into one end of it. Test to see that you can blow the whistle through the hose. Now twirl the dangling end of the hose with the whistle attached to it, at the same time blowing the whistle. The pitch of the whistle will rise as it moves toward your ear, and fall as it moves away.

Demonstrate *interference* by seeing how sound waves can cancel each other out. Get two equal lengths of rubber hose and a "T" tube, as shown. Attach the two hoses to the two ends of the "T." Place the two free ends of the hose against a watch on the table, then listen through the empty end of the "T" tube. You will hear strong ticks distinctly as the two sounds reinforce each other. Now begin to clip short lengths of rubber away from one of the hoses with a pair of scissors. The ticks will become faintest when the tubes differ in length just enough so that the push of one sound wave will cancel out the pull of another.

Try the "sympathetic milk bottles" experiment to illustrate *resonance*. Get two milk bottles just exactly alike and tell a friend to hold one up to her ear, as shown. Stand three or four feet away and blow sharply across the mouth of the other bottle until you produce a note. The air columns in your friend's bottle will pick up similar vibrations so that she will hear a faint note produced by *resonance*.

Demonstrate, by analogy, how sound waves disturb air molecules by doing the "marble molecule" experiment. Get six or eight marbles and suspend them on double strings on parallel supports as shown. You can make the marbles stick to the strings by blobs of gum or sealing wax. Have the suspended marbles touching each other and wait till they are perfectly still. Pull back on the first marble and let it swing against the next in line. The middle marbles will hardly move, while the end marble will swing out forcefully.

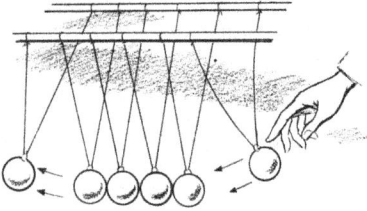

Watch how *overtones* are produced by doing this experiment. Remove the gong from an electric bell and mount it on a board as shown. Next attach a string to the clapper and carry the string over a hook or nail at the end of the board. Create tension on the string by attaching a weight to the end of the string over the hook. Start the clapper vibrating. As tension on the string varies, the string will vibrate in parts.

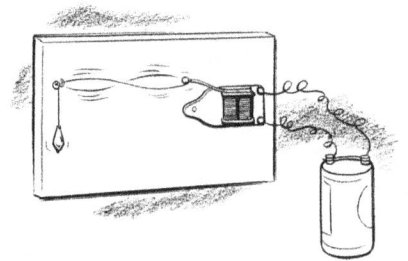

Make a "Pipes of Pan" by taking a strip of corrugated cardboard and shoving eight soda straws through every other opening of the cardboard. Blow across the tops of the straws to create notes of different pitch. "Tune" the straws so that you get a complete octave by clipping each successively shorter.

Index

www.ingramcontent.com/pod-product-compliance
Lightning Source LLC
Chambersburg PA
CBHW081110220326
41598CB00038B/7302